目

錄 Contents

自 序

一九八八年我隻身到台中求學時年僅十九，對於從來不曾離開台北的我來說，台中僅只是個陌生的地理名詞，因為念書的機會我才認識了台中，猶記得當時每到星期假日，我常隨著當地的同學玩遍大中部地區，從中橫的山川美景到彰化鹿港小鎮，以至西海岸養蚵人家，還有南投的田園風光等等，很多的自然美景頓時讓我真正的體會到為什麼書本上總是稱台灣這塊地方為「寶島」。

二年後，我從學校畢業入伍分發到金門服役，那是一個位在廈門灣裡的小島，在那裡我感受到一股與台灣不同的地方特色，長期沒有對外開放的戰地，保留了相當道地的風土民情，傳統的閩南建築、人文深深的吸引著我，使我開始想運用微薄的薪餉及休假回台的時間到民藝品店裡蒐集各式各樣耐人尋味的古老器物，從此開啟了我對民藝品蒐集的喜好。

在眾多的台灣民藝品中，我發現傳統手工製作的竹器是比較不被重視的，並且多數都沒有受到良好的保存，尤其是很多戶外使用的農、漁具等特別容易腐壞，加上六○年代政府大力推行農漁業現代化科技改良政策，使得很多傳統的竹製生產器具都遭到淘汰，而其他室內用的竹製品，也都被現代化的家電、塑膠製品等取代了，好比家家戶戶有了冰箱以後廚房裡再也不用竹製的菜櫥；客廳裡新的沙發也替代了舊時的竹椅，一時間大家棄竹器如敝屣，眼看著這些傳統美好的手工竹製用品就將隨著二十世紀一起走入歷史時，一股發自內在的情感促使我將這些殘存散落各地的竹器找回，並著手將這批竹器整理出版。

過去這近百年來，竹器曾經與台灣人民生活有過一段非常緊密的關係，早期的農村社會，人人過著簡單清苦的生活，日出而作，日落而息，生活中所使用的竹器都能相當程度地反映出台灣人民的思想與生活的習性，因此這些傳統手工製作的竹器在邁向新世紀的今日看來，理應被視為一種具有時代意義及台灣文化特色的文物來善加保存才是。

近兩年來，我有機會到歐洲及美國等地旅行，參觀當地著名的博物館，回台後，再次面對這一批在自己土地上孕育出來的舊時代器物，我的感觸更深，因為歐美等先進國家對於這樣的文化資產都非常的重視，並都做有系統的保存與推廣計畫，我們也該加緊做好才是，日後在地球村中，也讓世界各地的人來看看台灣先民生活的智慧及美德，我想這樣的保存與推廣才更有意義。

這本書的內容主要是以各式竹器圖片的登錄為主，文字介紹為輔，希望這本書的出版，能引起更多人對台灣竹器有更多的關注，共同來珍惜這項寶貴的文化資產，這是我編著此書的最大期望。

李贊壽 05/2000

Preface

In 1988, at the age of 19, I went to Taichung to study. I had never left Taipei before and Taichung, to me, was merely the name of an unknown place on a map. By going to school there, however, I got to know the area. I remember spending every vacation with my local classmates traveling around central Taiwan - taking in the gorgeous scenery of the central mountain range and the small town of Lugang in Changhua county, as well as visiting the oyster farmers of the west coast and pastoral scenery in Nantou county. The beautiful scenery I encountered let me discover for myself what is meant when books describe Taiwan as "Formosa" (a treasure island).

Two years later, I enrolled for military service and was sent to Kinmen, a small island in the bay off of Xiamen (Amoy). There I discovered a local flavor that differed from that which prevailed on Taiwan. Having long been an isolated military stronghold, Kinmen retained its rich local customs and traditional Minnan architecture. The culture deeply attracted me and I found that I was soon using my meager wages and trips back to Taipei to wander in stores selling folk art, collecting all manner of intriguing antiques. That is how my love of collecting folk art began.

I have found that among the folk arts of Taiwan, bamboo handicrafts are the most neglected and many have not been well preserved. Bamboo fishing and farm implements in particular are subject to neglect and decay. The demise of these implements was speeded by government measures in the 1960s to modernize the agricultural and fishing industries. Household objects of bamboo also fell into disuse through modernization as they were replaced by electronic and plastic appliances. As household after household acquired refrigerators, there was no more need for bamboo cupboards to store food. Living room sofas replaced old bamboo chairs. All at once, these bamboo objects were discarded like old shoes. The thought that these beautiful traditional handicrafts may go the way of the twentieth century and fade into history spurred me to collect the remaining works scattered across Taiwan, organize and publish them.

For the past 100 years, objects made of bamboo have been closely tied into the lives of the Taiwanese. In early rural society, life was simple. People made a poor but honest living, rising at dawn to work until dusk. The beliefs and practices of the Taiwanese people are reflected in the everyday bamboo objects they used. From our vantage point, at the cusp of a new century, these traditional bamboo handicrafts are the valuable resource full of historical and cultural significance that should be saved.

Over the past two years, I have had the opportunity to travel in Europe and the United States and visit their renowned museums. Upon my return to Taiwan, I felt an even deeper connection to the antique objects that sprang from my own culture. Advanced nations such as the United States and those in Europe hold their local cultural production in great regard and have carefully catalogued and preserved them. We must do the same so that in the global village of the future others can appreciate the knowledge and virtue of traditional Taiwanese society, thus giving meaning to the collecting and preservation of such cultural artifacts.

This book serves mainly as a catalogue illustrating a variety of bamboo handicrafts, with supplementary text explanations. I sincerely hope that the publication of this book can stir up interest in Taiwan's bamboo handicrafts and lead to a concerted effort to treasure these cultural artifacts.

Patrick Lee
May 2000

緒　論

李贊壽

台灣，這個最早被葡萄牙人稱頌爲　"Ilha Formosa"　「美麗之島」的地方，從十六世紀開始，經歷了荷蘭人與明鄭成功大量的招徠漢人開墾，一直到十九世紀清末，仍有大量的閩粵地區漢人不斷地移居來台，使得台灣的生活文化普遍沿襲了很多大陸的形式。但是長期以來，礙於台灣海峽的阻隔，來台的漢人移民群逐漸與本地原住民相互交流融合；再加上漢人移民群對台灣地理氣候變化的適應與調整，久而久之便形成了一種有別於大陸內地甚至閩粵沿海地區的台灣文化特色。

然而就在即將邁進二十世紀的一八九五年，一紙「馬關條約」將台灣這個以漢人爲主體的多族群、多語言社會割讓給日本，使得台灣生活文化起了新的改變，在日本五十年的統治下，台灣隨著明治維新的西化運動逐漸開始了現代化與文明化，最明顯的例子如日本政府在台推行的「放足」政策，即鼓勵台灣纏足婦女解掉纏足、「斷髮」政策，即針對男性要剪去前清遺留下來的辮子，還有「易服」政策，也就是改變服裝，由唐裝換穿西服，這些改變對當時的生活文化無不都是巨大的衝激。此外，在日據時期興建的幾座知名建築物像是總統府、台灣銀行、台北賓館等西式建築，也都與傳統舊式的中國建築截然不同，這些都象徵著二十世紀初期台灣生活文化的改變。

另一個台灣生活文化演變得最快、最多的時期要屬二次世界大戰結束後迄今這段時間。一九四五年戰爭結束後，日本投降撤出台灣，在戰後一片破敗、貧窮和蕭條中，國民政府繼起經營台灣，當時艱苦的生活如吃蕃薯簽、穿破衣、打赤腳等，是老一輩的台灣人怎麼也忘不了的。但隨後農民配合政府從事農業改革，大肆建設農村，並且以農業生產來培養工業發展，五十多年來，台灣由農業社會進步到工業社會、商業社會，時至今日的資訊社會，國民年所得從幾十美元大幅提昇到一萬多美元，這當中生活文化的變遷，決不是一個快字或多字所能形容。就像人們至今仍然詫異長久以來伴隨他們共同承擔生活勞苦，勤勉任勞的耕牛，如今在台灣多數的農田裡居然已經看不到了，起而代之的是冰冷的機械鐵牛，數千年來的農耕文化就這麼徹底轉變了，或許有人會說這不僅是台灣才有的改變吧，是的，全世界很多地方的文化都在改變，而且變得很快甚至到最後大家都變成了同一個模樣，美國人吃麥當勞台灣人也吃麥當勞，歐洲人穿ARMANI台灣人也穿ARMANI，MADE IN TAIWAN的產品更是風行天下，到處都有，換言之，台灣文化的某些部分已經變成世界文化，世界文化更可能隨時成爲台灣文化，這是資訊社會的結果，也是文化巨變的主因，當資訊快速地流通成爲世界大同時，過去某些只屬於單一地方獨有的特色，也就顯得格外珍貴了。回顧劇變的二十世紀，台灣生活文化當中特有的「竹器」就是一個明顯的例子。

曾經與台灣人民生活朝夕相伴、情感與共的竹製器具，其用途之廣泛從初生嬰兒睡眠用的搖籃到逐漸長大學坐的椅轎、乳母椅；從客廳用的桌椅家具到廚房用的菜櫥、蒸籠；從農耕用的秧苗籃、挑秧架到捕漁用的漁筌、魚簍；再到婚嫁祭祀中最不可少的各式盛籃禮器等等，每一項都是和人們生活緊密相關到缺一不可的地步，也是先民幾代以來透過生活經驗的累積傳承，結合智慧與本土豐沛的竹資源，用無數的雙手件件編製而

成的，更是同時代中全球少數地區特有的一種文化資產。但是隨著時代的進步，文化的演變，人們不自覺的將這些竹器一一棄捨，而改用千篇一律機械化的塑膠金屬製品來取代，直到現在竹器幾乎已完全退出人們生活的舞台。所以處在這即將消逝的二十世紀末，面對過去台灣住民這一百年來的生活竹器，我們著實應將它視為地球村中具有一定時代意義及地方特色的文物來愛惜才是。

竹資源的分佈與運用

放眼世界，竹子的種類雖有一千二百多種，但多數都集中生長於亞洲，尤其東南亞就佔了將近百分之九十，在中國大陸竹子多生長於長江以南地區，長江以北只有少數零星的分佈，而台灣則是遍佈全島，從低海拔的平地到三千公尺高的高山到處都有，主要是竹子天性怕冷又不耐旱，適合溫暖溼潤的環境，而台灣地處亞熱帶，氣候高溫多雨，正是竹子生長的最佳環境，所以竹材資源相當豐富，以種類上來區分，就固有的品種加上外來種共計約有六十餘種，普遍常見的有桂竹、麻竹、綠竹、刺竹、孟宗竹及長枝竹等，都是具有實用價值與觀賞功用的。

宋朝的大文豪蘇東坡曾說：「不可居無竹」，又說「無竹令人俗」，可見自古以來中國人就喜愛運用竹子來製作各種生活器具，我們從浙江省吳興縣出土的新石器時代文物中，一批兩百多件竹篾編製成的器物，即可應証。而台灣住民在竹材的運用上，更是得天獨厚，因為隨手取得容易，在農、漁、食、衣、住、行、育、樂等民生方面，幾乎發揮到了無所不用的境界，連雅堂在《台灣通史》中便提到「台灣竹工之巧者，為床、為几、為籃、為筐，日用之器，各地俱有。」由此記載可見竹器在台灣住民生活中是多麼地普遍廣泛，而這些製竹工藝最早是隨先民自閩粵一帶傳遞過來的，來台後多半成為農村婦女農閒時的副業，只有較大型的家具是出自男人的手藝，起初各個遷居部落所製作出來的竹器都還各有其特色，但後來因為女子出嫁到別的部落，將製竹的手藝也伴隨著傳去，漸漸地各地之間相為融合，分別也就愈來愈不那麼明顯。同時隨著時間的演變，傳統閩粵地區的色彩逐漸退去，使得台灣特色更形鮮明。

二十世紀台灣住民的生活型態仍是以農、漁業為主，日常生活及生產過程中，所需的器具繁多，一般都就地取材直接運用，其中竹製品佔了相當大的一部分，若依用途的不同可約略作以下的分類：

一、農具部分有掛在耕牛肩上的牛軛、防止耕牛吃掉作物的牛嘴籠、餵牛藥的餵食器、用來掘土整地的刈耙、汲水灌溉的戽斗、作脫穀用的連枷、除蟲用的蟲爪子、挑秧苗用的挑秧器、秧苗籃、另外像是畚箕、扁擔、米籮、米篩、鐮刀籃、茶簍、烘茶籠、各式採收籃、茱籃、生畜籠及大型的輾米土礱、儲存稻米的穀倉等。

二、漁業方面，有各式漁筌、漁笱、漁齁、魚簍、蝦籠、水蛙籠、釣竿、救生筒等等。

三、食的方面有吃飯的筷子、筷子籠、盛飯的飯匙、飯攄、笊籬、汲油勺、碗籃、蒸籃、粿床、蒸墊、氣死貓（吊籃）、茱櫥及桌罩等。

四、在衣的方面有竹笥 （衣箱）、竹衣櫥、農人穿的披篷、斗笠、竹傘、竹帽、婦女用的針線籃、什細籃等都是很好的竹製品。

五、住的方面，台灣中南部特有的竹管厝、屋樑柱、竹編的門扇、室內各種竹桌、竹椅、屏風、竹床、竹枕、竹夫人等竹家具。

六、行的方面主要是嫁娶用的竹轎子及溪流中行駛的竹排、竹筏、竹篙等。

最後有關育樂方面的像是嬰兒用的搖籃、乳母椅、乳母車、竹童玩，還有各式竹樂器像簫、笛、笙等皆是。

出類拔萃的民間藝術

台灣住民的生活竹藝品非但是既經濟又實惠，更是諸多民間工藝中出類拔萃的一項。一九四三年日本民藝學家柳宗悅曾經來到台灣作實地考察，他對於台灣民間的日用工藝多數抱持肯定的態度，尤其欣賞深植於鄉土生活中所自然產生的美感，對於台灣竹藝更給予高度的評價，從以下這段柳宗悅對台灣民間藝術的評斷摘要，即可清楚看出：「……余為尋找形象美術之美，曾經歷遊朝鮮，中國東北地區，獨不知台灣，惟憧憬而已，此次有機會訪問台灣，頗感興奮。台灣之歷史比日本與朝鮮為淺，僅僅不過三百年，且它的民藝來自福建或廣東，故可謂台灣本身並無古來之民藝品，但實際調查結果，卻發現甚多可觀之作品。舉例而言，如其竹加工品，因深入民眾生活中，故似乎被視為極平凡之物品，但若以客觀之立場觀之，實為可驚異之存在。在幾年前世界有名的德國建築家普魯農・達多（Bruno Taut）前來日本考查時，他最驚異者為廉價之竹椅，尤其看到台灣產簡樸之竹椅時，恍惚良久。達多所感魅力者，為所用材料──竹之素材美及設計構造之特異。又商工省聘來日本之著名工藝專家貝利安女士對在高島屋百貨店展覽會場，所展出之台產竹加工品也讚賞地說：『此種竹加工品若被介紹到西洋，必定掀起大轟動。』云云。」

另外，發行於一九四三年到一九四五年的《民俗台灣雜誌》，主編金關丈夫在〈民藝解說〉專欄中，也有一段對台灣竹製品讚許的記載：「……台灣的民俗藝品，就屬竹製品最為人們所津津樂道。台灣本島居民生活的必需品，舉凡房子、船、家具到日用品幾乎都離不開竹子，因此台灣人對於製竹手法的巧妙，以及對於竹子那份特殊的感情是沒有人能比得上的，台南的關廟莊就對此項傳統有著完整的保存。近年來，內地(日本)有見識的工藝家開始招聘台南的製竹工人，打算在內地興起生活用具的竹器工藝化，結果還頗受好評。由台灣人為內地家庭引進美麗的竹器是再合適也不過了，並且對台灣人來說也是一項快樂的工作。年輕一代的內地人就從竹器開始實施台灣生活的計畫，不也是很適當嗎？……」

金關丈夫也曾對上文中所提到的台南關廟莊，作實地考察並有以下著述：「……竹細工在台南關廟非常發達，從組織、工作場所、工作形態、製品、銷售組織等，世界上再也找不到這樣的工藝村。工藝村本來是人們想像中一種近於理想的東西，而實際世上有這樣的村落確實令人驚嘆。（日本曾經招聘工人

到這裡指導，可是最後還是本土的工藝吃香）工藝品原本是有形的東西不是在比勝負，尊重本省鄉土的作品風格才不致於弄巧成拙。……」

可見台灣竹器的發展在日據時期已經臻至成熟，當時從事竹器加工的可說極為普遍，一般村民的生計以水田耕作為主，養豬與竹子加工大部分成了婦女的副業，就技術來分，竹器工藝可分為「編製」與「鑿製」二種，前者如各式提籃，也就是一般婦女所常編做的，後者則多為大型家具，泰半是出自男人的手藝。在造型與線條的表現上，竹器質樸無華的鄉土氣息，不造作、不矯情，完全流露著人與自然環境融合的本質，不加設色的外表，保留原有的自然色澤，在幾經歲月的洗鍊後所產生的那種特殊溫潤色澤與觸感，常叫人愛不釋手。

呈現生活的智慧與美德

二十世紀台灣住民生活竹器最可貴之處，在於其蘊含了無數住民生活的智慧與美德。早期的農村社會，多數過著簡單清苦的生活，吃著粗茶淡飯，日出而作，日落而息，克勤克儉，生活中所使用的竹器都是簡樸的必需品，很能反映出當時先民的思想與生活的習性。

好比竹器中最有代表性的——禮籃，就十足反映了台灣民間禮俗中先民謙卑有禮的生活習性。禮籃，又稱盛籃，一般以桂竹精編而成，在各地的神明壽誕、中元普渡、祈安三獻、建醮祭拜或男女訂婚、結婚等都會被派上用場，主要是盛載禮品和祭品用，通常分為媒人籃、檳榔籃、炮籃、層籃及謝籃等。其中檳榔籃是一種小巧精緻的禮籃，用在婚禮中盛置檳榔招待客人，以取其諧音「相敬如賓」做為討喜之意，可惜現代流行的西式婚禮中，已缺少了這項傳統。另外台灣有句俚語說「提籃子假燒金」，是形容一個人裝模作樣，要掩飾本意的意思。原因是過去人們凡是要到廟裡上香禮佛，必定提著裝盛祭品的禮籃恭恭敬敬前去，所以提著禮籃與上廟燒香在過去是形同等意的，但以前常有未出閣的姑娘提著禮籃假裝要到廟裡燒香，其實本意是要外出會郎君的，所以先民都會打趣的以「提籃子假燒金」來調侃。記得筆者唸書的時候，例假日想出去玩，都會拿著書本向父母稟告說要去圖書館唸書，母親總是一眼就能看穿我的本意，冷冷地回我一句「提籃子假燒金」，而我也就心照不宣了。這句俚語就這樣普遍地流傳著，可是在塑膠袋充斥取代一切容器的今天，若不將傳統美好的竹編謝籃好好保留下來，誰還能從這句俚語中體悟到先民提著禮籃到廟裡燒香，那種虔誠恭敬的態度呢？

台灣還有一句俚語「母雞不蓋，毋怪老鷹」，意思是提醒人們凡事要反求諸己，自我反省。早期農村社會家家戶戶都會在自家院子裡，露天飼養一些雞、鴨、鵝等家禽，並且以一個像倒蓋碗形的竹編罩子，將體型較小的母雞或小雞罩住，以免遭到從天而降的老鷹抓走，這是村民普遍的習慣。但農忙時農婦卻經常疏忽忘記罩住，於是雞禽常常會被老鷹抓走，心疼不捨的農婦總是仰天咒罵得逞的老鷹，一旁開明的長者就會用這句俚

語「母雞不蓋，毋怪老鷹」來訓示村婦，要怪只能怪自己大意疏忽忘了將雞禽罩住，不能怪罪老鷹。在現代生活中，我們有時會看到社區裡幾戶鄰家的孩童玩著玩著，不一會兒就吵起架來，匆忙趕來的家長，愛兒心切總是護著自己的小孩，互指別人家的不是，卻不知要先自我檢討自我約束，想想先民這句俚語「母雞不蓋，毋怪老鷹」，自己家的小孩沒先約束好怎能怪別人呢？如今寸土寸金的都市，已少有人會奢侈到劃地露天養雞，而生態遭到破壞，老鷹也幾乎消失了，有誰還能記得這個長得像個倒蓋碗形的竹編罩子呢？這樣的文化資產你說是不是應該極力來保存呢？

另外，有一種竹器叫「氣死貓」，光聽它的名字就會讓人莞爾一笑。那是一件由竹篾編成的大型吊籃，在還沒有冰箱，物資匱乏的年代裡，被利用來儲放魚肉等食物，吊掛在廚房的樑柱上，以防鼠輩或貓兒偷吃，讓看得到卻吃不到的貓兒，只能在吊籃下踱步徘徊、望梅止渴，於是人們打趣地將這件吊籃稱為「氣死貓」。先民這種安貧樂命、幽默風趣的生活觀是多麼地值得我們學習，但反觀物慾過重的今天，九二一大地震後停電的那段時日，有一些人成天為了電力資源不足輪流限電而吵翻天，殊不知幸福的現代生活就算一天裡少個幾小時的電，也還是比以前人的日子好過太多了吧，可是現代人卻少了先民那種安貧樂命、幽默風趣的生活觀，大家大概都忘記了氣死貓長得甚麼樣子，也忘記了氣死貓的生活哲學，你說是不是很可惜？

時過境遷，睹物思情，竹器背後所隱藏的往事與意義不勝枚舉，對於舊時代先民惜福愛物的美德，我們要心生感佩，要加以傳承，因此保存與記錄這批竹器留給後人回想，誠屬迫切必要。

逐漸改觀消失的竹器

台灣住民生活竹器大抵在二十世紀初即逐漸有了自己的風貌，在形式與編製手法上，一步步走出閩粵地區移民的色彩，少了正規中原造形的束縛，也不再受講求華麗的牽絆，實用方便樸拙簡約的島嶼風采已然顯見。在日治五十年期間的後半段，也就是二〇年代到四〇年代之間，竹器已結合台灣住民的生活美學達到成熟發展的階段，從日常實用機能中散發出美感享受，因而才能得到日本學者的重視與肯定，在推展到日本及其他地區後大受歡迎。

台灣光復後，政府視竹藝加工品等為具有經濟利益之產業，遂成立「手工業推廣委員會」，大肆推廣手工業，一方面使戰爭期間散失的舊有技術人員重回生產行列，並增加就業人口，再方面藉由手工藝品的外銷增進外匯，一直到七〇年代左右，這項政策每年都為國家帶進可觀的外匯收入，只是外銷的手工藝品如竹篾編製的麵包籃、聖誕裝飾等，都是因應國外市場的需求，已經不具有台灣鄉土的特色了。同時由於農業政策的改革，政府一連串的「加速農村建設方案」、「促進農業全面機械化計畫」、「發展精緻農業」等措施，使得農村生活形態大幅改變，傳統的竹製器具不是被機械取代，就是因轉作而遭到擱棄，在一場鄉村都市化的變革中，過去美好的竹器一度被視

為落後的象徵，迅速地從人們生活的舞台上消失，塑膠製品隨後攻佔取代，民間工藝美學頓時湮滅無存。

直到八○年代後，人們緣於機械產品的單調無味，使生活陷入千篇一律枯燥空虛的氛圍，才又掀起一股懷念鄉土的民藝熱潮，紛紛回到民俗藝品店中撿拾已遭遺棄多時的器物，重溫過去手工藝產品所給予人們的自然觸感，再次享受它所散發的生命力量，以及美的韻律和喜悅。近年來，社會形態隨著一日千里的科技一變再變，獨特的台灣竹器風貌早已改觀消失，連最可貴的民情風俗都難耐新文明的衝激，雖然國人物質生活愈來愈加富裕，但精神生活卻大不如前，因此找回與繼承優良的文化傳統，實乃勢之所趨。

展望未來

回顧二十世紀，台灣住民生活竹器從極盛一時到銷聲匿跡，讓人不勝婉惜。而今要跨世紀將這項美好的傳統工藝來延續與發揚，當務之急首重保存與記錄，將周遭留存的竹器加以保養維護並集結出版，盡可能地完整記錄下每一件竹器的使用背景與文化意涵，使能永久流傳。其次是透過公開展覽介紹，喚起各界對竹器的重視與重新認定其價值。最後再落實回生活教育中，一面傳承竹器的製作與使用，另一方面要發揚其背後潛藏的優良傳統，最終達到提昇人文生活的內涵；及美化生活空間的目的。

本文僅止於對二十世紀台灣住民生活竹器作一概略之介紹，相關之深入議題仍有待你我共同來關心研究，一起為傳承美好的台灣竹器貢獻一分心力。

Introduction

Patrick Lee

Taiwan, called "Ilha Formosa" by the Portuguese, was first settled in the 16th century by the Dutch and then the Han Chinese who came with Ming resistance leader Zheng Cheng-gong. A steady flow of Chinese immigrants from Fujian and Guangdong provinces in southern China came to Taiwan up until the late Ching Dynasty in the 19th century, giving Taiwan a cultural make-up very similar to mainland China. Over time, given the barrier of the Taiwan Strait, a cultural synthesis was created between the Han Chinese and local aboriginals. As the Chinese immigrants adapted to the geographic terrain and climate of Taiwan, the culture developed characteristics distinct from those of the mainland interior and even those of coastal southern China.

In 1895, in accordance with the Treaty of Shimonoseki, Taiwan, with its Han-majority pluralistic society, was ceded to Japan, setting into motion even greater change for the population of the island. Over the course of Japan's 50-year rule, Taiwan was swept along with the Meiji reformational processes of westernization and modernization. The most noticeable changes occurred as a result of the Japanese policies to rid Taiwan of the old custom of footbinding, to encourage men to cut off the queues they had worn under the Ching, and to adopt Western styles of dress. These changes greatly affected the lives of the people in Taiwan. Additionally, the Japanese built several notable Western-style buildings, such as the Presidential Office, the Bank of Taiwan and the Taipei Guest House, which were a great departure from traditional Chinese architecture. These examples all provide a representative picture of the great changes gripping Taiwan in the beginning of the 20th century.

Another important era in the history of Taiwan came with the end of World War II. When the Japanese were defeated in 1945, they were forced to give up Taiwan, which at that stage was dilapidated, poor and desolate. It was under these conditions that the KMT took over control of the island. The older generation of Taiwanese will never be able to forget the hard times that prevailed when they had to subsist on sweet potatoes, wear ragged clothing and go barefoot.

The new government undertook agricultural reforms, integrating agricultural production with industrial development and bringing about the establishment of the ruiral area. Fifty years later, Taiwan's agricultural society was transformed into industrial and commercial society, which is now entering the information age. The personal income grew from tens to over US$10,000. This degree of change can hardly be quantified by terms such as "fast" or "great." To this day, people are still surprised to find that the diligent and uncomplaining oxen that accompanied them in the are rarely seen anymore. They have been replaced by machinery. In such a short span of time, the agricultural practices of thousands of years had been utterly transformed.

Of course, some people may be insist that these types of changes are not specific only to Taiwan. That is true. The entire world is changing - not only at a furious pace but also along the same model. American eat McDonald's, and so do the Taiwanese. European wear Armani, and so do the Taiwanese. The "Made in Taiwan" label can be found everywhere. Taiwanese culture, in many respects, is seeping into world culture, and vice versa. These great cultural changes can be traced to the information age. As the speed of information exchange creates an ever more homogenous world, old local characteristics become even more valuable. Taiwanese bamboo handicrafts are one such example.

These bamboo handicrafts, which were present in all aspects of Taiwanese daily life, had a multitude of uses. Accompanying Taiwanese from birth in the form of cradles, bamboo handicrafts evolved as children grew into forms such as chairs with protective bars used when the child is first learning to sit and "mother and child chairs," which could be used both by adults and children. In houses, bamboo handicrafts took the forms of chairs, tables and other furniture. In kitchens, there were bamboo food cupboards and steamers. Farmers used seedling baskets and carrying poles to cultivate the fields, while fishermen used bamboo fish traps and baskets. The ritual of marriage would be incomplete without the abundance of wedding baskets and ceremonial trappings so closely tied with daily life that to be missing just one would be unthinkable.

These were products passed on from generation to generation, fashioned and refined over time from each generation's accumulated experiences. The handicrafts combined local wisdom and copious natural resources of bamboo. Taiwan's bamboo handicrafts hold their own among other regional cultural handicrafts. However, with progress and cultural change, mass-produced, machine-made plastic and metal objects have replaced bamboo ones and today, bamboo handicrafts have almost retreated entirely from daily life in Taiwan. Thus, as we look back on the last century of Taiwanese life and the place of bamboo handicrafts in it, it is important that we value these handicrafts as cultural products of a global village with definite historical significance and regional characteristics.

The Spread and Use of Bamboo
There are over 1,200 varieties of bamboo, most of which are concentrated in Asia. Ninety percent of the bamboo in Asia is found in Southeast Asia. In China, bamboo grows mainly in areas south of the Yangzi River. North of the river, bamboo grows sparingly in scattered areas. Bamboo is spread in enormous quantities all over the island of Taiwan, from the low-lying coast to 3000-meter mountains. Since bamboo naturally thrives in warm, wet climates, the subtropical heat and rain of Taiwan make it the perfect environment for bamboo. As a result, there was no shortage of bamboo resources. Of varieties, Taiwan has over 60, including local and imported types. Most commonly used were guei zhu (phyllostachys makinoi), ma zhu (dendrocalamus latiflorus), lu zhu (bambusa oldhamii), ci zhu (b. stenostachya), mengzong zhu (p. pubescens), and changzhi zhu (b. dolichloclada).

Sung Dynasty the man of letters Su Dong-po said of bamboo that one could not live without it and that to lack bamboo was to be uncultured, showing the status of bamboo from ancient times to the present for everyday objects. The excavation of over 200 Neolithic artifacts woven from bamboo strips in Wuxing county, Zhejiang province is further proof. For Taiwan, particularly rich in bamboo resources, it was a part of every facet of life - farming, fishing, food, clothing, houses, transport, child-rearing and music, among others. According to Lian Ya-tang, in Taiwan Tong Shi Yi Shu (A General History of Taiwan, Book 1), skilled bamboo artisans made beds, tables, baskets and chests, and everyday objects of bamboo could be found everywhere. From this description, it is apparent that bamboo handicrafts were extremely common and widespread.

The earliest artisans followed the traditions brought over from southern China. The majority of the handicrafts were produced by the rural women and only larger pieces of furniture were made by men. Initially, each region produced its own style of bamboo handicrafts. As children married and moved to different regions, they brought with them their own region's styles, resulting in an eventual merging of styles. At the same time, the style of bamboo handicrafts drifted away from that of southern China and a distinctive Taiwanese style was born.

Twentieth-century Taiwanese life revolved around agriculture and fishing, which required a huge assortment of implements. Most of these were made with local resources, of which bamboo figured prominently. Implements could be divided according to their uses as follows:

1) Agricultural tools included yokes for oxen, muzzles, feeding tools, plows, irrigation devices, flails, bug catchers, seedling pickers, seedling baskets, baskets for carrying soil, poles for shouldering loads, rice baskets, rice sieves, sickle baskets, tea baskets, tea drying baskets, carrying baskets, vegetable baskets, animal cages, grinders and granaries.

2) For fishing, there were various fish traps and baskets, shrimp cages, frog cages, fishing poles and life-saver barrels.

3) For eating and cooking, chopsticks, chopstick containers, rice scoops and strainers, colanders, oil ladles, bowl containers, steamers, glutinous rice cake steamers, steamer sieves, a type of hanging basket called "infuriate the cat," food cupboards, and table covers were all used.

4) Under the category of clothing, items such as chests, cabinets, farmers' raincoats, rain hats, umbrellas, hats, women's sewing baskets and baskets for sundry items all represent fine examples of bamboo handicrafts.

5) In the central and southern regions of Taiwan, one can find bamboo huts, beams, doors, tables, chairs, screens, beds, pillows, and pillow-shaped arm and foot rests for sleeping in the summer.

6) Items used in transport consisted mainly of wedding sedan chairs and rafts and punting poles.

Bamboo handicrafts which played a part in child-rearing included cradles, baby chairs, baby carriages and toys. Instruments made of bamboo included the vertical flute, panpipe and flute.

A Remarkable Folk Art
Taiwan's bamboo handicrafts were not only economical and practical, they also stood out among the region's folk arts. In 1943, Japanese folk art scholar Yanagi Soetsu visited Taiwan to do research. His admiration for Taiwanese folk handicrafts was strong and he had a special appreciation for the beauty of the natural forms produced by rural villagers. His appraisal of Taiwans bamboo handicrafts was particularly high. The following passage, written by Yanagi Soetsu on Taiwanese folk art, clearly shows his regard:

> In my search for the embodiment of the beauty of art, I have in the past traveled to Korea and northeast China. I did not go to Taiwan, which was a place I could only imagine. This opportunity, then, to visit Taiwan had me quite excited. Taiwan's history does not approach the depth of Japan's or Korea's, being only 300 years. One could even say Taiwan itself has no ancient folk arts as they have all come from Fujian and Guangdong. However, upon completing my survey, I have discovered numerous worthy objects. Bamboo products are the example. One could regard them as extremely ordinary since they enjoy widespread use in everyday life. However, from an objective view, one can not help but marvel at them. Several years ago, when world-renowned German architect Bruno Taut (1880-1938) came to Japan, he was most impressed by low cost bamboo chairs and upon seeing the simple Taiwanese bamboo chairs, he was entranced. What Taut found most charming about the chairs was the material used - the beauty of bamboo and the unique designs and structures it inspired Madame Belianz, another folk art expert brought to Japan by the department of commerce, remarked of the Taiwanese bamboo handicrafts displayed at the Takashimaya

Department Store: "If these bamboo products were introduced in the West, they would create an enormous stir."

Kanaseki Takeo, the editor of Minsu Taiwan Zazhi, a magazine about Taiwanese folk customs published between 1943 and 1945, wrote a column on folk art, part of which also praises Taiwan's bamboo products:

> Of Taiwan's folk arts, people are most passionate about those made of bamboo. These objects are a necessity for life in Taiwan and everything from houses, boats, furniture and everyday objects could not do without bamboo. Therefore, Taiwanese artisans have developed exquisite skill in their production and the relationship the Taiwanese have with bamboo is unparalleled. The tradition is well preserved in the village of Guanmiao in Tainan. Over the past few years, experienced artisans in Japan have begun looking for bamboo craftsmen from Tainan in the hopes of creating a market for handcrafted bamboo objects in Japan. The results have been favorable. Having Taiwanese introduce Japanese households to these wonderful bamboo objects is only right, and furthermore, the Taiwanese find this type of work quite enjoyable. The younger generation of Japanese can use the bamboo crafts industry to launch their own plans in Taiwan.

After observing the village of Guanmiao, Kanaseki Takeo described it as follows:

> Bamboo craftsmanship in Tainan's Guanmiao village is very developed. Guanmiao is unique among villages producing handicrafts in terms of their organization, workshops, work attitude, products and marketing system. As these types of villages are often imagined idealistically, it is a surprise to encounter such a village in actuality. These handicrafts have their own unique form and attributes and do not need to be judged as if in a contest. The local style should be respected to avoid ruining a good thing. [Japan had sent workers over to direct to local artisans, however, traditional Taiwanese craftsmanship still ended up being the most popular.]

By the period of Japanese rule, the development of Taiwan's bamboo craftsmanship was already mature and the craft was very commonplace. The majority of rural villagers relied on fishing and agriculture. Women also raised pigs and produced bamboo handicrafts. Woven handicrafts, such as baskets, were made by women, while items requiring heavier labor, such as large pieces of furniture, were crafted by men. The modeling and lines of the handicrafts reflect the uncontrived, rustic flavor of the countryside and the close connection of the people with nature. Unhampered by fussy details, the handicrafts retain the natural hue and characteristics of the bamboo, which over years of use ripen to an appealing warm glow and texture that many find captivating.

Presenting colloquial wisdom and virtue
The greatest value of twentieth-century Taiwanese bamboo handicrafts is what they reveal about the colloquial wisdom and virtue of the people who used them. In early rural society, people led a simple, hard-working life. People made a poor but honest living. They were diligent and frugal, eating simple fare and rising at dawn and working until dusk. The simply designed, utilitarian bamboo objects they used are an excellent source for gleaning information about their beliefs and lifestyle.

The ceremonial basket most fully reflects the customs and beliefs

15

underlying ceremonial practices of earlier generations of Taiwanese. These baskets were finely crafted from guei zhu and usually contained gifts and offerings. They were an integral part of festivals for the birthdays of gods and deities, rituals for ghost month, offerings for peace, temple celebrations to commemorate anniversaries, and engagements and weddings. Some commonly seen examples are the matchmaker's basket, betel nut basket, fireworks basket, tiered basket and ceremonial basket. The exquisitely-made betel nut baskets were filled with betel nuts to treat wedding guests and signified a marital blessing that the couple would respect each other as they would guests. Unfortunately the modern predilection for Western-style weddings has brought about a decline in such traditional customs and implements.

A Taiwanese folk saying - "By carrying a basket, one pretends to make offerings to the gods" - describes people who hide their true intentions under the cover of a false alibi. The meaning derives from the practice of carrying baskets of offerings to the temple to worship the gods. Visits to the temple were frequent and the sight of people carrying ceremonial baskets on their way to the temple was common. Soon, carrying a ceremonial basket on the street became synonymous with going to the temple. Unmarried girls often used this pretext to arrange romantic liaisons. Carrying a basket of offerings, they were free to wander about as if on their way to the temple to piously burn incense.

I remember pulling the same trick when I was a student. On holidays, when I wanted to go out to play, I would grab a book and tell my parents I was going to the library to study. My mother always saw through me though and would coolly say, "By carrying a basket, one pretends to make offerings to the gods." This saying has been handed down to us from earlier generations and has embedded itself into our common parlance. However, with the ubiquity of plastic products, who will remember the way our ancestors used to carry these baskets to the temple if we do no make a concerted effort to save the traditional bamboo baskets?

Another Taiwanese saying - "You can't blame the hawk if you don't cover the chickens" - reminds people not to throw blame on others, without first examining themselves. Every household in the farming villages used to raise fowl, such as chickens, ducks and geese, in the yard. The women would feed them in the open air of the yard and in order to protect the chickens and small chicks from being carried away by hawks, the women would cover them with woven bamboo bowl-shaped covers. Often, in the busy planting or harvesting seasons, they would forget to cover the chickens. When the hawk swooped down and stole the chickens, the women would scream curses into the air after the retreating hawk. The village elders would use the saying to admonish the women, as they could only blame themselves for forgetting the cover the chickens in the first place.

A modern example of this pertinent saying lies in every neighborhood, where we often see children playing with the neighborhood kids. Inevitably, a fight breaks out among the children and each parent rushes to their own child's defense, each blaming the other parents for failing to properly restrain their children. But these parents fail to remember the wisdom of the saying, which reminds them to first look at themselves. If they have not properly taught their own children, how can they blame others for failing to do the same? In our modern cities, where time is money, people no longer feed chickens out in the yard and the danger of the hawk is but a memory. Under such circumstances, who now would remember the bamboo cover that was used protect the chickens? This is precisely why we must strive to preserve these cultural artifacts that provide the basis for our present culture.

Another type of bamboo basket is given a whimsical name that no doubt brings smiles to people's faces - "infuriate the cat." This was a large hanging basket, woven of bamboo strips. Before refrigerators, these baskets were used to store meat and fish and hung up on kitchen beams. Such contraptions frustrated the attempts of household vermin and cats to try and get to the food. With the food in clear sight, temptation was high, but the cat could only pace the floor below the basket and stare longingly at its contents. Thus the basket was given the clever and humorous name - "infuriate the cat."

Earlier generations of Taiwanese had a healthy attitude toward life that we would do well to learn from. Their content with the simple lives they led and their ability to apply humor and wit to the things around them are qualities we could use in our present overly-materialistic society. After the 921 earthquake, for example, people made a big fuss over the electricity outage. A day without electricity is too much to bear for some people, but what they forget is that even then, their life is far easier than that of their ancestors. What people today are lacking is the ability to be happy with what they have and see things with humor. Everyone has probably already forgotten what an "infuriate the cat" basket looks like, and has certainly forgotten the philosophy behind it..

Traditional bamboo handicrafts have the power to evoke a time gone by, holding within them a wealth of information about the past. The virtues of our ancestors, who led spare but contented lives, are worth not only our respect but also our efforts to carry on in their footsteps. Therefore preserving and recording these bamboo handicrafts for our generation and those of the future is an urgently necessary task.

The Gradual Disappearance of Bamboo Handicrafts
By the beginning of the 20th century, Taiwanese bamboo handicrafts were already coming into their own and stepping away from the styles that had been carried over from Fujian and Guangdong provinces. Straying from the mainland style of regular and restrained forms and rigidly ornate decorations, Taiwanese handicrafts focused on function and were notable for their rustic, pared down style. Towards the end of Japanese rule, between the 1920s and 1940s, Taiwanese craftsmanship had reached a mature stage of development. The beauty expressed in the art forms was derived from functionality, an aesthetic which garnered the appreciation and praise of Japanese scholars, who embraced Taiwanese folk arts and spread their popularity in Japan and elsewhere.

After Taiwan returned to Chinese rule, the new government, recognizing the value of the local handicrafts, set up a committee to oversee the advancement of these arts. They helped former artisans recover from the ravages of war and revive production and also actively recruited new members into the artisan community. The committee was also responsible for pushing the export of Taiwan's handicrafts, which brought in considerable revenue until the 1970s. The export products, however, had little to do with those originally made for use in rural village life. Following foreign tastes, products for export were mainly bread baskets and holiday decorations.

The government simultaneously enacted several measures to reform the agricultural sector. Policies to quicken the pace of rural development, mechanize agriculture and boost the sector resulted in a massive transformation of village life. As villages grew more and more like cities, traditional bamboo arts were left on the wayside. They came to symbolize backwardness and were all but wiped out by mechanization and plastics.

It wasn't until after 1980 that cold and sterile machine-made products devoid of meaning started to revive in people a nostalgia for village folk arts. People searched shops selling folk art to recapture the joy that these objects - imbued with life and the beautiful rhythm of nature - could provide. As our society is carried forward by the unrelenting speed of technological advancement, Taiwan's unique bamboo handicrafts have unfortunately long disappeared and folk customs and practices have already been overcome by modern civilization. Although our lives are richer in a material sense, we are spiritually depleted when compared to the past. As such, the valuable cultural traditions of the past are worth recapturing and passing on.

A View Towards the Future

Looking back on the twentieth century, from the proliferation of Taiwanese bamboo handicrafts to their extinction, provokes a profound sense of loss. It has thus become an urgent priority to conserve and record these beautiful traditional arts in order to ensure they are not truly lost to us forever. By preserving, protecting and publishing the remaining objects, along with explanations of how they were used and their cultural significance, it is hoped that they can be preserved for a long time to come. By bringing these handicrafts to the public's attention, there can be renewed interest and understanding of their value. It is hoped that the production and use of these handicrafts can be passed on as well, and that the traditions embedded within them can add to our present culture.

This essay is only an introduction to Taiwan's twentieth century bamboo handicrafts. Together we can all embark on a deeper understanding of these arts.

圖　版
Plates

家具篇

家具，是生活文化的重要構成者之一，一個地方或一個民族的生活文化，往往會在其所使用的家具風格中表露無遺，比如傳統中國的明式家具就充分展現出古時候中國人生活之中莊嚴、幽雅及高貴的氣息；同樣地，在我們研究一百年前台灣早期使用的家具時，會發現很多從福建、廣東沿襲過來的風格，顯露了當時移民的文化色彩；又，在稍後的日治五十年中，台灣家具明顯地受到日本西洋式風格的影響，逐漸擺脫了舊中原的形式，也象徵性的說明台灣生活文化改朝換代的更迭；今日，世界各國名牌家具大舉入侵台灣人的居家生活，一股新世紀、地球村的大同文化，就將在你我的身邊蔓延開來。

百年來，在台灣的農村家庭中，所普遍使用的竹製家具——竹桌、竹椅、竹床、竹櫃等，其經濟輕巧、簡單實用及可直、可折、可曲的特性，完全符合了那個物資缺乏、生活從簡，凡事要向大自然環境妥協適應的時代所須，也因此這些竹製家具普遍都有一種象徵著台灣人簡單樸實的獨特風貌，相對於移民時期仿清式或廣式的家具形制，或日治時期泛西洋風格的木製家具來說，竹製家具更能真切地反映出台灣住民的生活文化。

過去，要製作一件竹家具，也並不是那麼地容易，從竹材挑選、去節、除油、乾燥到加工成形，每一個環節都積聚了許多人的智慧和心血，一件作工仔細的竹家具，通常可以用上好幾代，只可惜在一昧追求現代化、西洋化的七〇年代，人們迫切地要在家中給電冰箱、電視機、洗衣機、沙發椅等找到一個位置，而不管這些沙發家具在氣候潮濕酷熱的台灣是否合用，就硬是將舊式的竹家具給搬走或丟掉，甚至於劈了當材燒，才會使得今天要找來一件完整堪用，又有點年代的竹家具，會如登泰山般的困難，所以，對現存僅有的要更加地珍惜，很可能，這已是保存竹文化資產的最後機會了。

Furniture

Furniture is an essential component of daily life and culture, and regional and ethnic culture is often exhibited through furniture styles. Ming furniture, for example, reflects the stately, elegant and noble tastes of ancient China. By examining the furniture used a century ago in Taiwan, we get a sense of the prevalent styles that came over from Fujian and Guangdong provinces. During the period of Japanese rule, Taiwanese furniture styles were influenced by Japan's fascination with the West. The change in furniture style, which gradually shifted away from its Chinese origins, also symbolizes the general cultural change Taiwan was experiencing at the time. Today, Taiwan is becoming a part of global culture as more and more designer furniture from around the world makes its way into homes.

A hundred years ago, the furniture seen in Taiwanese homes exemplified the unique simplicity and practicality of the local style. The finely-crafted, economical tables, chairs, beds and cabinets took advantage of the abundance of resources and were perfectly adapted to the rhythm of local life. The harmonious relationship of the people with nature was also evident in the furniture through the use of bamboo, a versatile material capable of adopting straight, bent or curved forms. This local style, in contrast to the styles of the early immigrants who imitated Qing or Guangdong furniture or the later Western styles produced under Japanese rule, best reflects the lifestyle and culture of Taiwan.

Making a piece of bamboo furniture is a rather complicated process. From choosing the bamboo, flattening the nodes, removing the oils, drying and actually working the bamboo, each piece of furniture involves the knowledge and sweat of several people. A well-crafted piece of bamboo furniture can be used for generations, but in the 1970s, people raced to keep up with modernization and westernization. Making room in their houses for refrigerators, televisions, washing machines and sofas, people got rid of their old bamboo furniture. Some pieces of furniture were even used for firewood. It did not matter that the sofas were less suited to the hot and humid climate than bamboo furniture. Now, trying to find antique pieces which are intact and can be used is nearly impossible. Those rare finds that do surface should be treasured as this could be the last opportunity to save these bamboo cultural artifacts.

001 靠背竹椅
Bamboo side chair　91×38×32 cm

竹製家具的造形全以線條構成，簡潔俐落的特色，在
這件竹椅上流露無遺。

The form of bamboo furniture is composed of linear
elements. The clean design and well-made quality
of this type of furniture are abundantly exhibited
in this chair.

002 靠背竹椅
Bamboo side chair　73×46×34 cm

這類較低矮的靠背竹椅，過去本省的民家幾乎戶戶都
有，輕巧便利，左鄰右舍常拎著這樣的竹椅，到戶外
聚集，或做家庭副業，或閒話家常。

In the past, nearly every household owned low side
chairs of this type. The chairs could be conveniently
carried out of the house for gatherings or gossiping
with neighbors or for doing chores.

003 竹製太師椅
Bamboo tai shi armchair　94×60×51 cm

本省民屋的正廳兩側慣常設有太師椅，以供接待賓客
或家中長輩，早年以閩粵師傅所製的木質太師椅為主，
此類竹製太師椅於日治中期以後始興起。

The chair of this type lined the main hall of a typical
Taiwanese home and were used by guests and senior
members of the family. Earlier chairs of this type
were made of wood following the style of craftsmen
from Fujian and Guangdong provinces, however,
craftsmen began to use bamboo during the middle of
Japanese rule.

004 竹製太師椅

Bamboo tai shi armchair　93×50×43 cm

竹製太師椅有一定的基本架構，只是扶手與靠背的裝飾線條常有變化，此件太師椅的靠背中央嵌有瓷片，兩旁搭以烘彎成S形的細竹稈，樣式非常好看。

Although armchairs adhere to a basic structure, each chair differs in the decoration and shape of the armrests and back. In this example, the splat is inlaid with porcelain and the side panels of the back are decorated with thin stalks of bamboo heated and bent to form S-shaped curves. These decorative details produce a highly pleasing effect.

005 斜背躺椅

Slant-back reclining chair　90×72×51 cm

這種有斜度的靠背躺椅，半坐半躺，甚為輕鬆自在，不若太師椅般拘謹，是工作後稍息片刻的良伴。

Chairs with slanted backs allow the sitter to partially recline, making them more relaxed than the rigid armchair. The reclining chair was used in moments of rest after one's work was done.

006 斜背躺椅

Slant-back reclining chair　85×74×51 cm

在六〇年代我國技術援外所派出的農耕隊中，竹工隊曾援助非洲、中南美洲等友邦栽種竹類，編製竹器，曾深獲好評，右圖這款休閒躺椅當時在異國即廣受歡迎。

In the 1960s, bamboo craftsmen joined agricultural teams in providing technical assistance to allies in Africa and Central and South America. Their efforts were trained on growing bamboo and producing bamboo products. Reclining chairs, such as the one pictured on the right, were very well received abroad.

007 **竹製太師椅**

Bamboo tai shi armchair 95×49×39 cm

這件太師椅的靠背及扶手面，全以烘彎成S形的細竹條組成，較為特殊少見。

The armrests and back of this armchair are composed of S-shaped bamboo strips that were heated and bent to create the curved lines. This design is quite unusual.

008 **靠背竹椅**

Bamboo side chair 84×37×34 cm

此件竹椅的結構與圖001相同，但細微彎轉處收歛較多，整體上更為樸實。

The structure of this chair is similar to the chair depicted in fig. 001. However, the curved elements of the chair are not as sharp, giving the chair a more rustic appearance.

009 **靠背竹椅**

Bamboo side chair 91×50×50 cm

此件靠背竹椅有竹稈彎製的弧形扶手，乘坐面也較一般寬大、舒適。

This side chair has arc-shaped armrests made of bamboo stalks. The seat is larger than most chairs of this type and more comfortable.

010 **竹製太師椅**

Bamboo tai shi armchair 82×50×42 cm

此件太師椅做工細膩，靠背的竹稈粗細一致，竹節整齊，樣式極為大方。

This armchair exhibits very fine workmanship. The back displays a consistency in the thickness of the bamboo stalks and the joints on each stalk are evenly aligned. The overall effect is natural and elegant.

011 **竹公婆椅一對**
Pair of bamboo gong po chairs　78×43×39 cm
公婆椅一般佈置在私人房內，與安置在廳堂的太師椅
不同，都有靠背而無扶手，四隻腳平直落地，樣式簡
單俐落。

These chairs were reserved for the private chambers of a
 home, unlike the armchairs which were placed in public
rooms. They are simple but well-made, with straight legs
and a back rest but no arms.

圖011之側面特寫
Side view of fig. 011.

012 **竹公婆椅一對**
Pair of bamboo gong po chairs　107×47×37 cm
這對公婆椅的椅背高長，坐靠時頭頸可倚在靠背橫桿
上，頗為舒適。
This pair of chairs has a high back which allows one to
rest the head and neck against the crest rail.

013　**雙人座竹椅**

Two-seat bamboo chair　92×82×48 cm

此件雙人座竹椅雖為單人座椅之加倍延伸，但整體造形及美感卻另有風味。

This two-seat chair is essentially a doubled version of chairs seating one, however the resulting form and aesthetic is quite different.

014　**延伸式躺椅**

Full recliner　168×66×56 cm

這類躺椅可坐可臥，在台灣的夏日午后，用來小睡片刻是再好不過的了。

This type of recliner could be used for both sitting and fully reclining. It is perfect for napping on a summer afternoon.

圖014延伸後之特寫。

Fig. 014 in full reclining position.

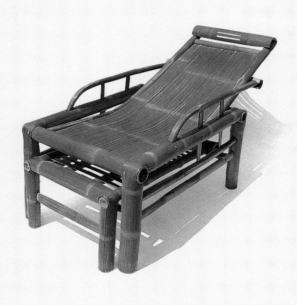

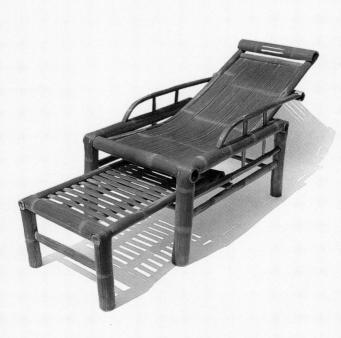

015　竹太師椅

Bamboo tai shi armchair　97×49×43 cm

現存成套的竹太師椅不多，往往不是缺几就是單只有一張椅，難得此組太師椅保存完好，整組的太師椅更能呈現完整的民俗風味。

Complete set of armchairs are now rare. The table that accompanied the two chairs is missing or only one of the chairs in the set remains. Therefore, this set of armchairs is noteworthy. The complete set also more fully illustrates style of the period.

016　茶几

Tea table　72×36×33 cm

通常置於兩椅之間，用來擺放茶杯或佈置花器，有讓兩椅的坐者能保持適度間隔的作用，而不會感到壓迫。

This type of table was usually placed between the two armchairs and was used for cups of tea or flowers. It also created a comfortable space between the occupants of the two seats.

圖015，016 成套二椅一几之特寫。

Full set of two chairs and tea table comprised of fig. 015 and 016.

017　竹太師椅

Bamboo tai shi armchair

（椅 Chair）　96×50×43 cm

（几 Table）　80×42×31 cm

這是另一套二椅一几的竹太師椅，椅背嵌有瓷片，茶几設有格層，保存相當完善。

This chair is from another set of two chairs and one table. The back of the chair is inlaid with porcelain and the table has multiple tiers. The set has been preserved in excellent condition.

018　剌竹凳
Ci zhu stool　26×26×19 cm

這是最簡單的竹凳架構，四方體各邊以一剌竹稈垂直接
合，造型簡潔有力。

The frame of this stool is simple designed. The four
rectangular sides are made of bamboo stalks connected at
perpendicular angles. The form is succinct and powerful.

019　竹凳
Bamboo stools

（1）31×31×27 cm　　（2）32×32×24 cm
（3）58×29×23 cm　　（4）54×29×26 cm

這類高低不一的竹凳，大人小孩通用，輕巧容易搬移，
在過去廣為人們愛用。

These stools of assorted heights were used by adults and
children alike. They were cleverly crafted, portable and
very widely used.

020　花架（三合一）
Flower stands (set of three)

（大 Large）　67×33×33 cm
（中 Medium）　45×25×25 cm
（小 Small）　28×19×19 cm

三件一組的花架，專用於擺放盆栽。

The three-stand-set is used to hold flower pots.

圖020三件花架大、中、小分開的特寫。

Separated stands of the set in fig. 020.

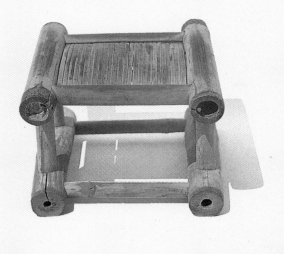

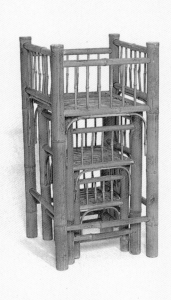

021 刺竹凳（四件一組）

Ci zhu stools (set of four)　38×31×31 cm

刺竹質密肉厚，堅固耐用，是製作家具及做為建材的上好材料，此類竹凳往往能用上數十年不壞，過去台南關廟一代即以出產刺竹家具聞名。

The flesh of ci zhu bamboo is dense and thick, making it a firm and durable material for making furniture and buildings. This type of stool could be used for upwards of ten years without breaking. In the past, Guanmiao in Tainan was famous for making ci zhu furniture.

---

022 刺竹桌

Ci zhu table　107×70×67 cm

製作刺竹家具比一般竹家具困難並且費時，無論劈剖、鑿洞、打竹釘都需要嫻熟有力的技術，非一般細竹工或學徒輕易可為。

Working with ci zhu to make furniture was more difficult and time-consuming than using other types of bamboo. All stages of the process - paring, chiseling, making pegs - required a high level of skill and experience. Therefore, ordinary craftsmen and apprentices could not easily produce these pieces.

圖021，022合成一套之特寫。

Fig. 021 and 022 combined as a set.

---

023 刺竹桌

Ci zhu table　92×60×50 cm

這類的刺竹桌，在北部較少人用，南部因為是產地所在，因此較為普遍。

This kind of table was seldom used in the north. It was much more common in the southern Taiwan where is the production center.

021 ◀

022 ◀

022.1 ◀

023 ▼

024 臉盆架
Washbasin stand　120×46×38 cm

這是一件保存的相當完整，年代及色澤極佳的竹製臉
盆架，構造極為巧緻，盆架有六隻腳，呈六邊形落地，
可擺放臉盆、香皂盒及披掛毛巾，非常牢靠實用。

This stand has been preserved very well. It is an early
piece with beautiful hue. Very delicate constructed, it has
six legs in hexagonal formation and can accommodate the
washbasin, soap container and towel. It is a very sturdy
and practical piece of furniture.

025 竹層架
Bamboo shelf　91×85×46 cm

簡單的三層結構，沒有繁縟的花樣，但卻滿足使用者
調理空間的需求，是一件很好的竹家具。

The simple three-tiered construction is not decorated but
fulfilled its purpose of ordering and organizing a space.

026 竹櫃子
Bamboo cabinet　63×61×31 cm

看似簡單的竹櫃子，做工卻極為綿密，作者巧妙的在
門扇的竹條上，削出四個菱形的空格，非但有通風的
功效，更具有裝飾的美感。

This bamboo cabinet looks simple but with quite finely
detailed. Four diamond-shaped ventilation slits have been
carved into the strips of bamboo on the panels of the doors.
They are beautiful decoration and serving practical purpose.

027 **竹床**

Bamboo bed 177×67×54 cm

這是一張供作休憩片刻用的單人竹床，床面以桂竹劈開的竹條組成，平滑舒適，一端並設有竹枕，樣式十分雅緻。這樣大型的竹製家具，過去因爲太佔空間，不好保存，多數都被劈成柴燒，因此還能留存到今日的，件件皆很可貴。

This single bed was used for a nap. The top is made of split strips of smooth guei zhu. At the end of the bed is a bamboo pillow. Large pieces of furniture like this were difficult to preserve since they took up a lot of space. Many pieces ended up as kindling and what has survived until today is considered highly valuable.

在看過這些竹家具後，不禁要讓人折服，同樣的幾支竹桿和竹條在先民的智慧巧手中，竟能變化出這麼多樣的器具，真叫人感佩。

The simple stalks and strips of bamboo were transformed into lots of variety of different amazing products. The furniture made from the wisdom of our predecessors are deeply admired and appreciated by us.

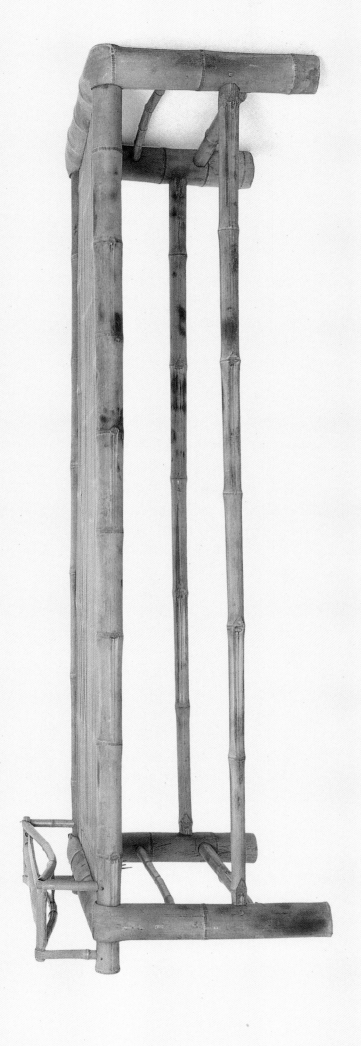

農具篇

台灣農業在近百年來雖然經歷了「農業台灣，工業日本」的殖民擠壓，但也在日治下大興水利，改良稻作，奠下深厚的基礎，戰後國民政府大力推展「三七五減租」、「公地放領」、「耕者有其田」、「加速農村建設方案」、「發展精緻農業」等等措施，使得台灣農業發展一時登上了頂峰，間接帶動社會繁榮，創下舉世稱羨的經濟奇蹟，然，任何人皆無法相信，曾經輝煌燦爛、盛極一時的台灣農業，竟然在完成「以農業培養工業」的使命後，從七〇年代起，日漸衰竭，終成為一種夕陽產業。

隨著農業的凋敝，農地廢耕，農村人口大量外移，這些陪伴著農民一起歷經風霜，數十寒暑，打下農業江山的農具，如今竟遭任憑遺棄腐朽，這樣的晚景淒涼，看在勤儉惜物、辛苦耕作了一輩子的老農民眼裡，不禁有無限感慨，俗話說「吃果子拜樹頭，吃白米拜鋤頭」，這種飲水思源的美德，現代人實在不應該視之淡薄。

在過去的台灣農村生活中，婦女們除了要在家相夫教子外，農忙時節也得到田裡去幫忙，同時還得兼顧種菜、撿柴、捕魚捉蝦等事兒，真是一刻都不得閒，工作量可以說一點都不比男人少。另外，每個家庭總都會飼養一些家畜、家禽來撿食三餐的剩飯或廚餘，這些家畜、家禽可不是現代人拿來把玩當寵物的名貴貓、狗，而是豬、牛、羊及雞、鴨、鵝等之類，不光是餵食、清理舍寮，有時還得幫豬隻洗澡，照顧起來可得另費一番心思。

通常飼養這些家畜都不是為了自家日常食用，多半是等牠們長大後賣來貼補家用，應付孩子學校註冊的學費等，有些時候家庭裡同時有四、五個小孩在上學，每到寒暑假開學前，持家的母親都得為這一筆可觀的學費盤算好要賣掉那些家畜，並且提早飼養。

而雞、鴨、鵝等家禽雖說賣了也無濟於貼補家用，但逢年過節祭拜神明或兒媳坐月子、冬令進補等，可就少不了，因此飼養這些家畜、家禽所使用的竹編器具，對現在身為老婆婆或老太太的以前農村婦女來說，內心裡可真是充滿了無限熟稔的情感，新時代的職業婦女恐怕是無法感受的。

本篇所收錄的竹製農具，全是先民運用本省豐沛的竹材，結合經驗與智慧所做成的，每一件都是辛勤與勞苦的象徵，也是農民篳路藍縷的精神所在，睹物思情，除了念舊與喚起對農村的美好記憶以外，也期待你我都能共同來尊重這些文化遺產，善盡保存與維護的責任，承繼優良的傳統美德，延續好的文化。

Agricultural Tools

Although Taiwan was experienced to providing the agricultural accompaniment to Japan's industrial development during the period of Japanese occupation, large-scale water conservancy and improvements in rice cultivation laid deep foundations for the sector. After the war, the KMT enacted measures to reduce rents, lease public land, give cultivators land, speed the development of rural villages and boost the agricultural sector. These measures led the agricultural sector to its highest point, brought overall prosperity and were generally heeded as an economic miracle. At the time, no one would have thought that such a flourishing industry would go into decline in the 1970s following industrialization efforts.

With the decline of the agricultural industry, fields were left fallow and villagers left for the cities. The implements that accompanied farmers through hardship and countless chill winters and hot summers were left behind to rot. To old farmers, who have led a life of diligence and hard labor, these precious tools invoke a flood of memories and typify the folk saying that when one eats fruit, one must thank the tree, and when one eats rice, one must thank the hoe. That is, we must acknowledge the origins of what we enjoy, wisdom that should not be dismissed lightly.

The women of farming villages had their work cut out for them. Aside from managing the house and teaching the children, in the busy seasons, women would go out into the fields to help. They were also responsible for planting vegetables, collecting firewood, and catching fish and shrimp. Their work load was by no means any lighter than that of the men. In addition to all of their other duties, women also raised livestock. These animals were nothing like the cats and dogs people today keep as pets, instead they consisted of pigs, cows, sheep, and fowl such as chickens, ducks and geese. The women gathered leftovers from meals and kitchen scraps to feed the animals, kept their pens clean and sometimes helped to wash the pigs.

The animals were usually sold to supplement the household income when they matured and were not meant to be eaten by the family. The money was often used to pay for the children's school tuition and there were sometimes four or five children in school at the same time. Before the beginning of every school term, mothers would calculate the money they needed, select which animals were to be sold, and begin feeding them early.

Although raising fowl did not add much to the household income, they were indispensable for holiday feasts or when the diet needed to be enriched for the approach of winter or for a new mother the month following the birth of a child. The bamboo implements used to raise these animals and fowl tap deep into the emotions of old women who spent their younger days in farming communities. Such identification would be lost on today's career women.

The agricultural implements gathered in this book have all been produced from local materials from the accumulated knowledge and experience. They symbolize diligence and hard labor and embody the pioneering spirit of the early settlers of Taiwan. Aside from cherishing the past and fond memories of rural village life, it is hoped that we can develop a new respect for these cultural artifacts and preserve and pass on traditional virtues for the enrichment of our present culture.

028 竹笠

Bamboo hat　37×34×14 cm

農人遮陽蔽雨最愛用的莫過於竹笠，竹笠的樣式編法，各地有異，此款竹笠多風行於本省中部。

Farmers used bamboo hats to shield them from the sun and rain. Each region had its own style and method of weaving. This style of hat was popular in the central region of Taiwan.

029 竹笠

Bamboo hat　42×42×16 cm

這款竹笠南洋風味濃厚，很有可能是漁民自外地所傳入。

This hat with the strong Southeast Asian style. It was probably brought from abroad by fishermen.

030 西嶼竹笠

Western Island bamboo hat　47×47×18 cm

曾經盛行於澎湖西嶼島上，編工獨特精巧，與一般竹笠極為不同，二、三十年前，澎湖縣政府曾將其列為重要特產，可惜因製作費時，編工後繼無人，不多久即已失傳，現已成為民藝界難求的珍品。

These hats were popular on Penghu's Western Island and very different from other bamboo hats. The weave is unique and exquisitely executed. Twenty or thirty years ago, the Penghu county government listed these hats as a special regional product. However, as it was time consuming to make them, eventually no one continued the trade and it died out. Today these hats are considered rare and valuable by collectors.

西嶼竹笠全貌。

Overall Western Island bamboo hat

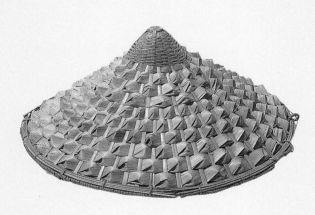

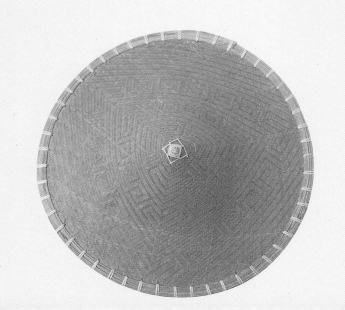

031 **龜形披篷**
Tortoise-shaped cover   85×76×22 cm
龜形披篷又叫龜甲笠，是因應農人插秧或除草時的特
殊姿勢需求所設計的，既防日晒又免雨淋，其上的四
個稜角正合背脊及臀部的角度，邊緣向外下垂，可使
雨水順勢流下，是一項具有生活智慧的聰明農具。

These were also known as tortoise-shell covers and were
used to keep away the sun and rain. They were designed to
suit the posture of farmers as they collected rice seedlings
or weeded their fields. The four corners follow the contours
of the back and shoulders. The outer edge flares out
directing rain water off of the cover. This clever item is
imbued with practical, everyday wisdom.

---

032 **貝形披篷**
Shell-shaped cover   85×76×22 cm
此為另一形式的龜甲笠，是用雙層竹篾條以網狀編組
而成，中間夾以麻竹葉或棕櫚葉梢，有防水之效，但
塑膠雨衣出現後，這類竹編披篷即已消失無蹤，現今
在各種竹編器物中，此類龜甲笠要算是最稀少難得的。

This cover is a variation of the tortoise-shell cover. It
consists of two layers of bamboo strips woven in a mesh
pattern. The inside is lined with leaves from ma zhu
bamboo or the tips of palm leaves, which make the cover
waterproof. Bamboo woven covers disappeared with the
introduction of plastic rainwear and among bamboo woven
handicrafts, these domed covers are now the rarest and
hardest to find.

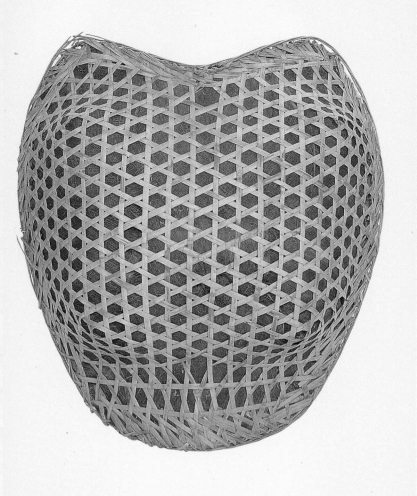

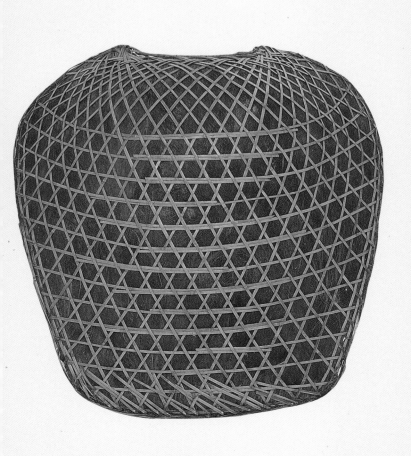

033　鐮刀籃

Sickle basket　27×17×15 cm

鐮刀是農人割草、割稻的利器，鐮刀籃是專用來存放
鐮刀的竹籃。

These baskets were for carrying the sickles used for cutting
grass and rice stalks.

034　蓮花籃

Lotus-shaped basket　43×43×14 cm

功用同圖035秧披。

For description of use see fig. 035.

035　秧披

Rice seedling basket　40×40×14 cm

插秧時用來裝秧苗的容器。

When planting rice, these were used to hold the seedlings
and shoots.

036　秧挑

Seedling carrier　64×21×3 cm

插秧時，本省南部多以此種秧挑來挑運秧苗。

These carriers were used for planting rice in southern
Taiwan.

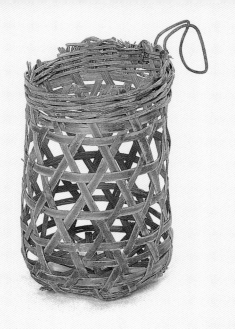

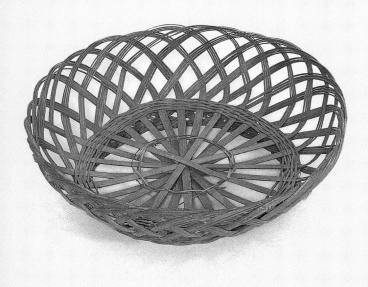

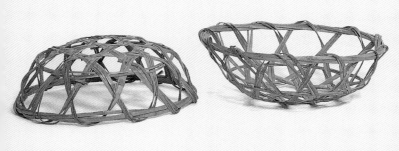

037 **鐮刀籃**

Sickle basket　23×20×20 cm

通常掛在打穀機上，放置鐮刀用。

This type of basket, used to contain sickles, was usually
hung on threshers.

---

038 **鑱箕**

Scoop　50×42×21 cm

盛稻穀專用。

These were used to hold paddy.

---

039 **簸箕**

Sieve　56×47×21 cm

作揚米去糠用。

These were used in removing rice husks.

---

040 **畚箕**

Baskets

（1）21×19×8 cm　　（2）37×34×17 cm

（3）28×24×11 cm

畚箕的用途多樣，拾牛糞、裝蕃薯、挑沙石等都用
畚箕，如圖（1）這類小巧無把手的畚箕，現已罕
見。

These baskets had many uses: to pick up cow dung, hold
sweet potatoes or carry earth, for example. This lovely,
handleless basket, such as (1), is now quite rare.

041　**蟲梳**

Insect comb　　183×34×33 cm

又名蟲爪子，農藥未興用以前，農人以此器來回橫掃
於稻葉間，將蟲子梳落在細竹片上再予以捕殺。

Before pesticides were used, farmers relied on these insect
combs to catch and destroy pests. They swept them
through the leaves and stalks of plants, catching the
insects in the fine teeth of the comb.

042　**除蟲器**

Insect catcher　　228×31×25 cm

功用與操作方式同上圖041蟲梳，唯本圖形式多爲本
省客家農莊所用。

The use of this tool is similar to that of fig. 041. This type
was used in Hakka farming villages.

043　**耙子**

Rake　　113×32×14 cm

用於曬穀時，可將夾雜在稻穗間的草葉碎枝耙出。

This rake was used to remove broken leaves and stalks in
the rice paddies.

044　**連枷**

Flail　　161×23×7 cm

爲一簡便的稻穀脫殼器，持連枷摔打於稻穗或大豆等
穀物上，穀殼即被擊脫。

These were simple tools used to remove rice and grain
husks. The paddy or beans were thrashed with the flail.

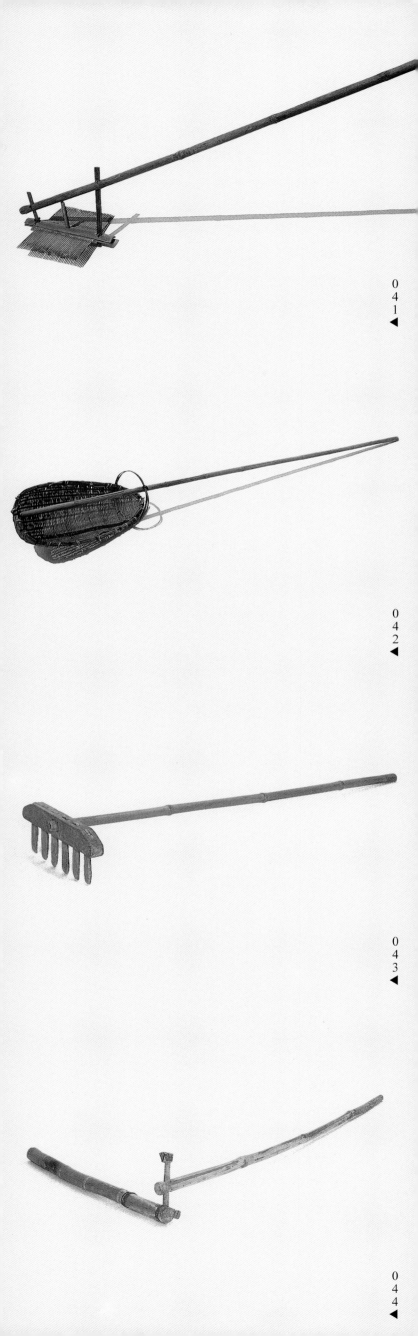

045 戽斗
Irrigation bucket　112×30×30 cm

汲水灌溉用具，專用於水渠難到之高田，現都以抽水機代替。

This irrigation device was used to carry water to elevated fields which could not be reached by irrigation ditches. Today, these devices have been replaced by water pumps.

046 竹提把水桶
Water pail with bamboo handle　40×32×32 cm

早期的木製水桶皆以烘彎的竹片作爲提把，後來便陸續被鉛鐵及塑膠製水桶給取代。

Early water pails were made of wood with curved bamboo handles. Later they were replaced by pails made of metal or plastic.

047 竹刈耙
Bamboo plow　106×76×24 cm

刈耙是用來切碎泥塊的農具，使用時以兩條粗繩由牛隻拉行，農人將雙腿叉開，一前一後踩在刈耙上，利用體重壓住刈耙下的竹刀，切開泥塊。

The plow was used to break chunks of soil. It was attached to oxen with two thick lengths of rope. The farmer would step on the plow, one foot on the front, one on the back, as it was pulled by the oxen and use his weight to push the plow's bamboo spikes into the ground.

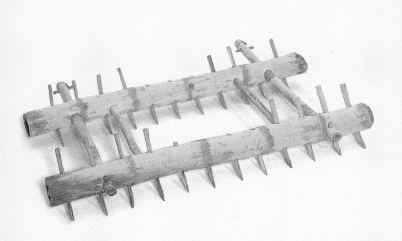

048 **茶葉簍**

Tea leaf basket 52×51×38 cm

此類茶簍使用時，採茶者通常是以麻繩穿透簍口，繫在腰側，然後將所採的茶菁塞入簍內。

This type of basket was used to pick the tea leaves. A rope was threaded through the vessel's mouth and tied around a person's waist. The person could then easily place picked leaves inside the basket.

---

049 **烘茶籠**

Tea drying basket 61×55×55 cm

採收下的茶菁經發酵、揉捻後，須再以此烘茶籠烘焙，以去除多餘水份。

After the tea leaves were picked, fermented, rolled and kneaded, they were dried by this basket to get rid of any remaining moisture.

---

050 **烘茶籠**

Tea drying basket 84×70×70

烘茶籠內編有凸起如倒置的漏斗形夾層，用以隔開茶葉與底部之炭火，中心點作透氣孔用，使用時將茶葉倒入烘茶籠之上部後，底下燒火烘焙，烘焙的時間長短可控制茶葉生、熟的程度。

The inside of this basket is divided in the middle by an inverted funnel. This serves to separate the tea leaves from the charcoal used to bake them. The tea leaves are placed in the upper half of the basket, with the charcoal underneath. The middle opening of the funnel allows for air to circulate. There is no woven bottom to the basket.The length of time controls the degree to which the tea is baked or raw.

（1）為圖050之底部特寫

View of bottom portion of fig. 050.

（2）為圖050之上部特寫

View of top portion of fig. 050.

(1)   (2)

051 **茶篩**

Flat tea basket　106×106×3 cm

茶農用此來裝盛烘焙後的茶葉，從中挑揀出不必要的茶枝。

Dried tea was spread out in these baskets and undesirable leaves were then plucked out.

---

052 **茶篩**

Flat tea basket　77×77×3 cm

有時婦人也會在茶篩上晾曬筍乾或瓜類作物。

Women also used this type of basket to dry bamboo shoots or melons.

---

053 **捻茶篩**

Tea kneading basket　42×42×8 cm

捻茶篩形為淺盆狀、軟邊，與一般茶篩呈平面狀、硬邊不同，尺寸上也比茶篩小的多，是茶農揉捻茶菁專用的。

These baskets are shaped like shallow basins and have soft edges unlike the flat tea baskets which are hard. The measurements of the kneading baskets are much smaller. They were used for kneading and rolling tea leaves.

---

054 **篩仔**

Shallow basket　58×58×10 cm

篩仔的功用主要在收穫穀物時，拿來剔除空殼及雜物，或用來曬種子，放豆鼓、年糕等，年節到來時，還可以在上面作粿或搓湯圓。

These baskets were used to pick out empty husks and debris when harvesting grains. They could also be used to dry seeds or hold food such as fermented soybeans or glutinous rice cakes. With the arrival of the new year, they were often used for glutinous rice or sticky rice balls.

055 **簳湖**
Large shallow basket　168×168×20 cm
一般的簳仔直徑大小都在二台尺左右，少數直徑超過
五台尺以上的叫做簳湖，專作曬穀用，此件即是。
The diameter of most shallow baskets is about 60 cm,
very few exceed 150 cm. Baskets such as this one, with
diameters over 150 cm, were used to dry grains.

---

056 **大篩**
Large sieve　70×67×13 cm
大篩又叫粟篩，呈淺盆形，全器編孔洞狀與簳仔密實
的編法不同，主要用在篩選稻穀、玉米等作物。
This implement, also called a grain or paddy sieve, is
shaped like a shallow basin. Unlike the dense weave of the
previous shallow baskets, it is loosely woven with holes. It
was used to sift paddy grains and corn.

---

057 **米篩**
Rice sieve　54×54×8 cm
米篩的孔洞邊長作物。
The holes of this rice sieve measure only 0.4 cm long, half
the size of the 0.8-cm holes of the large sieve. It has a flat
bottom. It was grasped in both hands and shaken back and
forth to sift rice or small green beans.

---

058 **篩子**
Sieves　（1）55×55×14 cm　（2）48×30×13 cm
（3）42×41×10 cm
此類篩子多用作篩洗煤炭或晾曬稻穀以外的雜糧。
These sieves were used for purposes such as washing coal
or drying rice paddy.

059 **米籮**
Rice luo basket   60×60×44 cm
籮是一種上圓底方的竹容器，多用以淘米或裝盛穀物，
此類慣稱做米籮。
Luo containers have round tops and square bottoms. Since
they were used for washing rice or holding grain they
became known as rice luo baskets.

米籮多成對使用，裝盛穀物後以扁擔挑行。
The baskets could be attached to poles for carrying.

060 **籮筐**
Luo kuang basket   43×43×28 cm
此種籮筐規模較一般米籮小，較多為室內使所用。
This basket is also a type of luo basket and therefore round
at the top and square on the bottom. It is smaller than the
similarly shaped rice luo baskets and was more often used
indoors.

061 **點心擔**
Food carrier   40×40×29 cm
此類點心擔顧名思義是農婦用來挑送點心給田裡工作
的男人用的，也有做生意時用來挑物品的。
This item is so called because it was used by women to
carry food to the men in the fields. It was also used to carry
products to sell.

062 牛嘴籠

Muzzle　20×20×16 cm

將牛嘴籠套在牛嘴上，可防止牛隻在耕作時損及作物。

The muzzle prevented the oxen from destroying the produce as they worked.

063 牛軛

Yoke　65×7×7 cm

又叫牛擔，一般多以刺竹稈的基部彎成，耕田、拉牛時跨放在牛肩上，再以繩索繫拖住犁即可。

Yokes were generally made by bending the base of ci zhu stalks. They were placed on the oxen's shoulders and secured with rope.

064 竹籮

Bamboo luo basket　20×20×20 cm

過去的農婦勤奮多藝，竹細工幾乎是必會的副業，像右圖這類的竹籮是很多婦人信手能編的，相當實用。

In the past, women took on a variety of duties. An essential one was producing small bamboo handicrafts. A basket such as the one pictured at right was something nearly every women knew how to make.

065 竹籮

Bamboo luo basket　23×21×21 cm

類似的竹籮樣貌很多，用途不一，所以各地名稱也多有差異。

Although many luo baskets look very similar, their uses are not necessarily the same. Each region therefore had its own name for the baskets.

066 **菜籃**

Vegetable basket   42×35×20 cm

這類竹籃常被用來買菜裝盛，一般慣稱爲菜籃。

This type of basket was used when buying vegetables.

067 **竹籮**

Bamboo luo basket   32×24×23 cm

這件竹籮的橢圓形編口緣造型是較爲特殊少見的。

The oval mouth of this basket is a special characteristic and
not commonly seen.

068 **竹籮**

Bamboo luo basket   29×29×24 cm

看似無奇的竹籮常蘊涵無數農業時代的舊回憶。

Though ordinary in appearance, these baskets remind us the
memories of the age of agriculture.

069 **竹籮**

Bamboo luo basket   30×21×21 cm

這件竹籮的篾條刻意削去部分的竹青，使表皮富有變
化，更爲美觀。

The outer skin of the bamboo has been partially pared away,
creating an aesthetically pleasing contrast.

070 **橘子籮**

Orange luo basket　34×24×24 cm

繫於腰間，採收橘子用。

The basket was tied to the waist and used for picking mandarin oranges.

071 **竹籮子**

Bamboo luo basket　44×29×29 cm

裝盛一般農作果物用。

This basket was used to hold fruit.

072 **竹籮**

Bamboo luo basket　29×17×17 cm

此件竹籮編製緊密，是專用來裝放較細粒的作物或茶葉茉等。

This tightly woven basket was used for holding smaller, finer products or tea leaves.

073 **雜糧籠**

Basket for sundries　26×26×22 cm

編製這件雜糧籠的竹篾粗厚堅硬，適於裝放雞蛋及雜
糧。

This basket is woven with thick, sturdy strips of bamboo,
making it suitable for holding items such as eggs.

---

074 **雜糧籠**

Basket for sundries　37×34×26 cm

專放甘藷類雜糧。

This basket was used for sweet potatoes.

---

075 **大竹簍**

Large bamboo luo basket　52×50×48 cm

供裝質輕量大的木炭使用。

This basket was used for large but fairly light items such as
wood and coal.

076 **雞籠**

Chicken cage　50×50×45 cm

此籠是以竹篾編成空格狀，並設有一門，側看略呈五角形，專供母雞撫育小雞用。

The cage includes a door and viewed from the side is pentagonal. It was used to keep hens with small chicks.

077 **雞籠**

Chicken cage　50×50×33 cm

此種雞籠是專供腳踏車載送待售土雞所用的，底平，上設有蓋，一籠有時可擠進十來隻。

These types of cages were used to transport chickens to the market on a bicycle. They have a flat bottom and a lid. Sometimes ten chickens could be stuffed into one cage.

078 **豬籠**

Pig cage　74×50×38 cm

豬販運裝小豬專用的。

This type of cage was used to hold small pigs for transporting them to sell.

079 **雞籠**

Chicken cage　38×38×32 cm

這是裝較大隻的雞或鴨，作客時當伴手禮用。

This type of cage was meant to hold larger chickens or ducks. Guests would use this kind of cage to bring presents to the host.

080 **小雞籠**

Chick cage　28×28×20 cm

裝雛雞或雛鴨用的。

These were used to hold chicks or ducklings.

081 **雞罩**

Chicken cover　78×75×41 cm

此種雞罩可將雞、鴨罩在固定地方，使其不會走失，也可防止雞、鴨到處便溺。

Covers like this kept chickens and ducks in one place and prevented them wandering away. It also prevented them from defecating everywhere.

082 **鵝籠**

Goose basket　64×32×30 cm

長形的鵝籠是專為長頸的鵝隻所設計的。

These long baskets were designed to accommodate the long necks of geese.

079 ◀

080 ◀

081 ◀

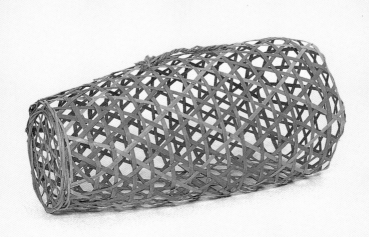

082 ◀

漁具篇

這裡所收錄由竹子製成的各式魚筌、魚苗簍、蝦籠、水蛙籠等漁具，泰半都是過去農民使用於溪河或田溝間的獵捕工具，倒非真正漁民用在海洋漁業上的，以前的農民除了耕種本業以外，早晚到鄰近的田溝或溪澗捕撈魚、蝦、鱔、鰻、水蛙及螃蟹等是經常的事，一來可多添加飯桌上的菜餚，再來也當做是農忙後生活中的一種休閒娛樂，但在七○年代工商業發展以後，許多天然資源遭到工業公害的污染，寶島的好山好水陸續受到工業區、發電廠、遊樂場、垃圾場等侵蝕，大大地改變了整個鄉間的生態，使得溪裡的魚蝦、田野的青蛙大為減少，有些種類甚至幾近絕種，因此，這些竹編漁具也失去了用武之地。

竹材編製的漁具，因為輕巧、便宜，加上竹材可彎的特性及易剖裂成粗細不同的竹條以利捕獲魚蝦等優點，成為農漁民的最愛，在農漁民巧妙的構思中，這些漁具個個別具造型與用途，例如最常見的捕魚竹器——魚筌，俗名「魚隔仔」，造型就有喇叭型、燈罩型、雞蛋型、花生型等多樣，大部分都以桂竹編成，口闊頸小，腹大身長，喉內編以放射狀的細竹，魚兒順著水流進入籠內，要再逆向游出就會受到喉內的細竹阻擋而不能復出，這類設計的魚筌最常用來捕吳郭魚、鯽魚、鱔魚及蝦；另一種捕魚或釣魚的人都要佩帶的魚籠——魚卡仔，也是以細竹篾編成，形體呈頸小腹大狀，頸部可穿上繩子，使用時繫於腰間，專門用來裝置漁獲，魚蝦一旦放進籠後就不易跳出。

另外，有一種漁具——魚苗簍，專門用來裝魚苗的簍子，是所有竹編器物中，唯一能盛水不漏的，原因是整個簍子除以細竹條密編外，在竹編空隙處另有填上牛糞及其它特殊塗料的混合物，所以可達防漏功效，這種魚苗簍通常編有四耳，穿上繩子後，漁民會選用一種特別細韌的扁擔來挑運，如此，在走動時，扁擔一上一下的彈動，就可使簍內的水不斷充氣，魚苗就不會因缺氧而死，這樣的生活智慧，怎不叫人由衷感佩。

Fishing Implements

The fishing implements recorded here - fish traps, fish baskets, shrimp and frog cages - were used by farmers in streams, rivers and the ditches among their fields, not by fishermen out at sea. When they weren't working in the fields, farmers would go out to catch fish, shrimp, eels and frogs, not only to supplement their dinner but also as a form of relaxation. The development of the manufacturing sector in the 1970s, however, led to the pollution of many of the island's natural resources. The mountains and waterways of "treasure island" gradually came under attack from factories, power stations, tourist parks and dump sites, completely changing the village ecology. The fish and frog populations declined and some varieties nearly became extinct. With them went bamboo fishing implements.

Fishing implements made of bamboo were well-loved for their convenience and low cost. The characteristics of bamboo, which could be worked into curved shapes and pared into thick or thin strips for weaving a variety of tools, made it well-suited for catching fish. Implements were made in a large number of different forms, each with specific functions. The most common was a type of fish trap which was made of guei zhu and could be horn-shaped, lamp cover-shaped, egg-shaped or peanut-shaped. The trap had a narrow mouth and wide, long body. Inside the mouth, sharp, thin spear-like pieces of bamboo faced inward. After fish swam into the trap, they were prevented from swimming out by the sharp spikes. This type of trap was used most often to catch mouth breeders, gold carp, eels and shrimp.

A fish cage made of thin strips of bamboo was also an essential piece of fishing gear. The cage had a small neck and a large body. A rope could be threaded through the neck of the cage, allowing it to be tied around a person's waist. It was used to hold the day's catch and the shape of the cage made it difficult for fish and shrimp to jump out.

Another type of fishing basket was used for holding minnows. It was distinctive because it was capable of holding water. Tightly woven with fine strips of bamboo, the basket was further coated with a paste made of cow dung, making it waterproof. The basket had four loops through which a rope could be threaded. It was then attached to a slender, pliable pole so that when it was carried the pole would move up and down with the person's gait. The movement served to aerate the water in the basket, keeping the minnows from suffocating.

083 **魚篊**

Fish catcher　53×34×34 cm

這類用竹篾編成像圍籬樣子的漁具，名為「篊」，使用時直接將篊插在水裡，把魚群罩住，再從篊的上方圓孔處一一將魚兒取出即可。

This device looks like a fence. To catch fish, it was thrust into the water and the fish trapped inside were pulled out through the hole in the top.

---

084 **魚笙**

Fish trap　（1）57×35×29 cm　（2）79×50×40cm

這類捕魚的竹籠，體形大大小小，有很多種尺寸，適合不同地形水域，過去普遍使用於本省中南部各地，北部較少。

This type of fish trap was produced in all sorts of sizes to suit different fishing grounds. It was more commonly used in the central and southern regions of the island.

右圖是上述魚笙之側寫，一端編著二道倒插的細竹，可防止魚兒逆游逃脫，另一端則是收取漁獲的開口，先民的智慧巧思，由此可見。

At right is the side view of the trap. There are two rows of overlapping bamboo strips at one end of the trap to prevent fish from escaping. The other end of the trap can be opened to retrieve the fish. This device is a good example of the ingenuity of Taiwanese early settlers.

---

085 **小魚笙**

Small fish trap　20×7×7 cm

這大概是所有魚笙中規格最小的一種，僅一手掌般大。

This palm-sized trap is probably the smallest variety of this type of fish trap.

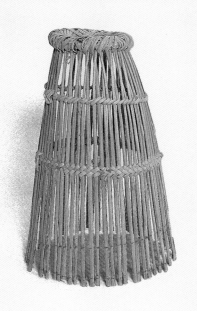

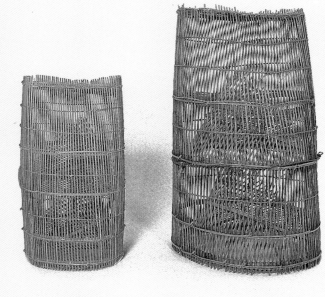

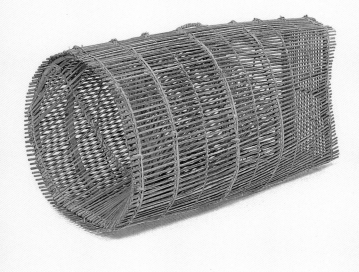

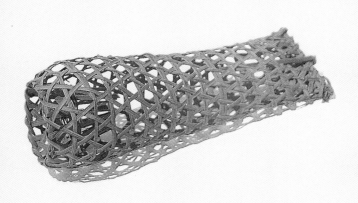

086　魚筌

Fish trap　56×44×44 cm

此類魚筌多以捕捉螃蟹為主。

This type of trap was mainly used to catch crabs.

087　魚筌

Fish trap　76×76×76 cm

此件開口特別寬闊，應是為廣納順流而下的魚群所設。

Traps with such wide mouths were probably used for
catching groups of fish as they swam with the current.

088　魚筌

Fish trap　57×33×33 cm

魚筌的形體大小和使用環境有密切相關，編者有時也
會依不同的漁獲量，來設定魚筌的形體大小。

The size of the fish trap depended on where it was meant to
e used. The size of the catch is also considered by the
craftsmen when making the trap.

089　魚筌

Fish trap　115×80×70 cm

此件魚筌無論樣式或規模都是眾多魚筌中所罕見的，
詳細的使用背景還有待做進一步的採訪調查。

Both the size and style of this trap are unusual. More
research is needed to fully understand its use and
background.

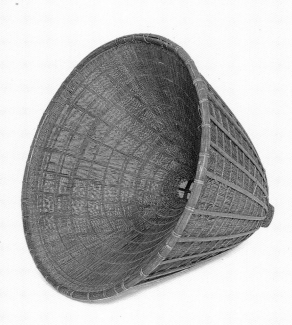

090　**魚筌**

Fish trap　48×20×17 cm

製作魚筌的另一個重要考量因素，就是得瞭解魚群的生活習性，如此方能有效發揮「請君入甕」的功能。

The most important factor to consider when designing a trap is the way the fish behaves. This way a trap can be designed that utilizes the fish's own natural impulses to catch it.

091　**魚筌**

Fish trap　70×16×10 cm

此種長柱形魚筌為捕鰻魚專用的，器身編有一孔洞，應為放餌所設。

Long, tubular traps were used to catch eels. Bait was placed in the hole in the middle of the trap.

092　**魚筌**

Fish trap　72×20×20 cm

此類魚筌多用於本省客家村莊，如桃、竹、苗一帶。

This type of trap was used by Hakka communities such as those in Taoyuan, Hsinchu and Miaoli.

093　**魚筌**

Fish trap　95×26×22 cm

此件形似花生，兩側都設有閘口，是眾多魚筌中較為特殊的。

This peanut-shaped trap stands out from among the other traps. Its two lobes each have an opening.

圖091之立面特寫及較小尺寸之同形魚筌
View of fig. 091 stood on its end with smaller trap of the
same style.

094 **碗形編籃**
Bowl-like basket 78×78×29 cm
漁販展售漁獲的容器
This type of basket was used for displaying fish when
selling.

095 **魚苗簍**
Minnow basket 36×36×36 cm
裝魚苗的竹簍子，因四周塗有牛糞等混合物，所以能夠
盛水不漏。
A cow dung and waterproofing paste was applied to
minnow baskets so that they could hold water.

093.1 ◀

094 ◀

095 ◀

096　魚苗撈

Minnow strainer　75×66×54 cm

這種魚苗撈編織緊密，滲水的細縫僅0.3公分，是專用來趕撈幼小魚苗的。

This type of strainer was tightly woven with holes measuring only 0.3 cm and was used to scoop minnows out of the water.

魚苗撈的尾端編有一漏口，用來倒洩魚苗，但在撈用時則另以一竹節頭塞住。

The hole in the back of the strainer was used to pour the minnows out after the water had been drained away. While straining, the hole was plugged with a piece of bamboo.

圖096魚苗撈的側面特寫。

Side view of fig. 096.

蝦籠

Shrimp cages 26×10×10 cm

此類蝦籠是專為捕溪蝦所用，使用時蝦籠內須放進魚餌，往天黑前將籠以石塊壓放在溪石澗，隔日清晨再去收籠，都會有不錯的收穫，而且每次可同時擺放幾十個這樣的蝦籠一把捕捉，如圖中這一擔就有三十幾個蝦籠。過去，全台各地都普遍使用這種蝦籠，而今溪河污染嚴重，還可以派上用場的地方，所剩寥寥。

These cages were used to catch shrimp in streams and creeks. Bait was placed inside each trap and they were laid out in streams right before dusk and anchored with stones. The next morning at dawn they could be retrieved and usually yielded a good catch. The advantage of these cages is that tens of them could be placed out at one time. Over thirty are pictured here. These cages used to be prevalent all over Taiwan, however, with streams and rivers now polluted, there are few places where they can still be used.

098 魚簍

Fish basket　45×29×28 cm

用來裝魚蝦的竹簍，開口設有漏斗型的倒插竹條，可防
止魚蝦躍出。

Fish and shrimp could be placed in this type of basket. Its
funnel-shaped opening, made of bamboo strips, prevented
the fish from leaping out.

圖098開口之特寫。

View of opening of fig. 098.

099 釣魚箱

Fishing basket　36×34×24 cm

多數釣魚的人都備有這樣的釣魚箱，共有三層，可裝釣
具及漁獲，十分方便。

These were carried by many people to hold fishing gear and
also the day's catch.

100 水蛙欄

Frog basket　32×31×30 cm

以鐵線穿綁竹條所編成的四方欄，造型簡單，是過去在
水田間釣水蛙，裝水蛙用的。

These simple, rectangular baskets were made of bamboo
strips bound together with iron and were used to put frogs
caught in the rice paddies.

0
9
8
◀

0
9
8
.
1
◀

0
9
9
◀

1
0
0
◀

101 魚簍

Fish basket 25×24×15 cm

裝魚蝦用，又名「卡仔」。

These were used to hold fish and shrimp.

102 魚簍

Fish basket 28×27×13 cm

本省各地所編的「卡仔」形式稍異，此為較常見的一種。

Each region's fish baskets had slight variations in form. The type pictured here was more commonly seen.

103 魚簍

Fish basket 35×30×20 cm

此款魚簍體形較大，可裝十餘斤的魚穫。

This style of fish basket was larger and could hold over 6 kg.

104 魚簍

Fish basket 50×30×23 cm

此類魚簍多半用繩子穿綁，以肩背或繫於腰間，視個人習慣而定。

These baskets were usually threaded with rope and either slung over the back or tied around the waist.

105 **魚簍**

Fish basket　18×16×11 cm

此款型式較小，多掛於腰間使用。

Smaller fish baskets were usually hung around the waist.

---

106 **魚簍**

Fish basket　21×18×18 cm

口闊、頸小、腹大是這類魚簍的共通特色。

The special characteristics of this type of basket are the
wide mouth, narrow neck and large body.

---

107 **魚簍**

Fish basket　23×21×18 cm

此類魚簍的外形，就像釣魚人的本性一樣簡樸。

This type of basket reflects the simple nature of the
fishermen who used them.

---

108 **魚簍**

Fish basket　23×23×23 cm

這類魚簍與大多數同年代的民藝品，如陶甕、錫器等皆
有著相通的美感。

This type of basket shares the same aesthetic qualities as
other folk art of the period, including ceramic jars and metal
wares.

109　**魚簍**

Fish basket　42×42×36 cm

這件魚簍以裝捕獲的鰻魚為主，圓形寬大的底部正合鰻魚彎綣的身軀。

This type of basket was used to hold eels as the large, circular bottom is well-suited to the eel's body when coiled.

110　**魚簍**

Fish basket　27×25×25 cm

此款為撿拾貝類所用。

This style of basket was used for gathering shellfish.

111　**魚簍**

Fish basket　36×36×32 cm

此款多用於河床邊，裝盛淘洗後的蛤蜊。

This type of basket was used to hold oysters after washing them by the riverbank.

112　**魚苗簍**

Minnow basket　42×42×30 cm

形若淺甕，是早期裝魚苗所用。

Shaped like a shallow jar, this is an early example of a container for minnows.

提 籃 篇

當竹器盛行的時代，竹工藝品店曾遍及全台所有鄉鎮，在各地都可以發現從事此種行業的蹤跡，根據台灣前輩畫家，也是工藝指導家——顏水龍先生的著述，竹工藝品行業的祖師爺，相傳是漢朝一位治理交趾地區政務的人——馬援，他首先運用當地豐沛的竹材製作成兵舍裡所須的各種器具，才使得後來竹工藝業漸漸興起，所以今天從事此行業的人便推崇馬援為祖師爺，並且各自在店裡供奉馬爺的雕像，來祈求生意興隆。

在六〇年代塑膠製品尚未普及以前，這些販賣竹製品的商店，每一家店裡所賣的竹藝品目都十分繁多，光是竹籃子一類就琳瑯滿目，有菜籃、花籃、秤籃、麵線籃、雞蛋籃、賣蚵籃、洗衣籃等多的不勝枚舉，這些籃子每一個都是手工編成的，無論形式、大小都不相同，有深也有淺，全恃編者的巧思及用途不同而定，不過，有時同一種籃子可有多種用途，因此在名稱上就不一而定，可能東家說是麵線籃，西家說是花生籃，所以，大凡用竹條編成而有提梁的盛物容器，都可以統稱為提籃。

而今這些前人所遺留下來的各式竹籃子，雖然已經完全退出人們現實生活的舞台，但在另一個文化資產的領域裡，它那一身可親可近的古早鄉情，卻一點一滴地沁浸你我的心扉，在不斷變遷的台灣社會中，成為早昔台灣人民勤儉性情與質樸形貌的永恆象徵。

Hand Baskets

Traces of the past apex of the bamboo handicraft industry can be seen all over Taiwan, where once craft shops dotted every town and village. According to Yan Shui-long, a painter and crafts instructor, the forefather of the art of bamboo handicrafts was a Han dynasty official in the district of Jiaozhi named Ma Yuan, who used bamboo to make items of daily use for the soldiers' barracks. To this day, artisans worship Ma Yuan as the father of the trade and make offerings to images of Ma Yuan in their shops to attract fortune.

In the 1960s, before plastic had made significant inroads in Taiwan, stores selling bamboo crafts had a vast selection of products. In terms of baskets alone, the variety was mind-boggling - there were baskets for laundry, vegetables, flowers, scales, noodles, eggs, and oysters to name a few. Each was entirely hand-crafted and variations existed among the shapes, sizes and depths of the baskets depending on the artisan and the purpose of the basket. Some of the baskets were used for more than one purpose and therefore went by different names. What was called a noodle basket to some, may have been known as a peanut basket to others. To simplify things for the purposes of this book, containers woven from bamboo strips and which have handles can all be categorized under the term hand baskets.

Although they can no longer be considered daily use objects, they can be seen as cultural artifacts which provide a close look at early village life in Taiwan. As Taiwanese society continues to transform itself, these artifacts symbolize the diligence and simplicity of a time gone by.

113　提籃

Hand basket　42×34×34 cm

此款提籃編工精整、牢固，從前婦人多用來裝衣物，提至河邊清洗。

This style of hand basket is superbly crafted and sturdy. Women used these baskets to carry clothing to the riverbanks to wash.

圖113手把的特寫，線條形式充滿美感。

Detail of fig. 116 showing the elegant lines of the handle.

114　提籃

Hand basket　50×35×35 cm

另一款婦人洗衣常用的提籃，編工綿密漂亮。

This is another style of laundry basket beautifully made with tight weave.

115　提籃

Hand basket　30×20×20 cm

此件提籃以較硬的竹篾編成，籃身紮實、形式美觀。

This basket was woven with stiffer bamboo strips and has a very pleasing shape and weave.

116 **菜擔**
Vegetable basket   63×34×34 cm
此類提籃多成對使用，以菜販挑運蔬果用最多。
This type of basket was used mainly by vendors
transporting fruits and vegetables to sell.

117 **提籃**
Hand basket   56×30×28 cm
刻意去除部分竹青後的篾條，常能使竹器的外貌更富變
化與美觀，此件即是。
By paring away part of the outer skin of bamboo, the artisan
creates a beautiful contrasting design.

118 **菜擔**
Vegetable basket   60×41×41 cm
另一種形式的菜擔，常見於昔日農村，是婦人挑菜沿街
叫賣用的。
This style of basket was used by women selling vegetables
on the street where they would call out to buyers.

116

117

118

119　**秤籃**

Weighing basket　46×46×46 cm

從前菜販秤重用的秤籃。

This type of basket was used by vegetable vendors to weigh their produce.

120　**提籃**

Hand basket　30×30×30 cm

此款提籃外形像蓮花瓣，又叫蓮花籃，從前婦人多作洗衣籃用。

This type of basket is also known as a lotus basket since it is shaped like lotus petals. It was used by women as a laundry basket.

121　**提籃**

Hand basket　44×40×39 cm

這類提籃多做菜籃使用，有時也會用來裝盛牛吃的草料。

This type of basket was used for vegetables, but sometimes also filled with grass to feed the cows.

122　**秤籃**

Weighing basket　31×27×24 cm

這是蚵販秤蚵仔所用的提籃。

This type of basket was used to weigh oysters.

119 ◀

120 ◀

121 ◀

122 ◀

123　**香花籃**

Flower basket　32×27×18 cm

編工精緻的香花籃，是從前大家閨秀用來掛放香花的。

This refined basket was used by girls from well-to-do families to hold fragrant flowers.

124　**提籃**

Hand basket　26×26×26 cm

多為提麵線用，所以也叫麵線籃。

As it was used to carry noodles, this type of basket was also known as a noodle basket.

125　**提籃**

Hand basket　28×24×22 cm

形同謝籃，無蓋，或做裝檳榔用。

Shaped like a wedding basket but without a lid, this may have been used for betel nut.

126　**小花籃**

Small flower basket　17×17×12 cm

上書有「中華民國三十四年　游金蓮」。

This basket is inscribed "The 34th year of the Republic of China (1945) You, Jin-lian."

127 **提籃**

Hand basket　32×23×20 cm

以竹篾做「人」字形的編法，是很多提籃所常用的，本件即是。

The bamboo strips are woven into a pattern commonly on many hand baskets. The pattern looks like the Chinese character "ren" (human).

---

128 **提籃**

Hand basket　25×20×20 cm

另一種「三角孔」編法，也是編製提籃所常用的，如本件即是。

Another type of commonly seen patterned weave for hand baskets is characterized by triangular holes.

---

129 **蚵仔籃**

Oyster basket　20×15×14 cm

這是海邊養蚵人家裝盛蚵肉用的提籃，所以叫蚵仔籃，籃身雖編得細密，但仍能瀝去多餘的水分。

This type of basket was used by coastal oyster farmers for holding oyster meat. Although it is tightly woven, the basket still allows excess water to drain out.

---

130 **蚵仔籃**

Oyster basket　29×22×22 cm

容量較大的另一型蚵仔籃。

This style of oyster basket has a bigger capacity.

131 **提籃**

Hand basket　36×26×22 cm

此類提籃纖細、輕巧的模樣，常會令人愛不釋手。

It is impossible not to love such delicate and nimble baskets such as this one.

---

132 **提籃**

Hand basket　50×39×39 cm

編製提籃所用的篾條，劈剝不易，老師傅常說的「劈剝三年」就是指要劈剝出寬幅、厚薄一致的篾條是需要長時間的學習與練習的。

Making the strips that are used to weave baskets is not easy. Seasoned craftsmen often say one must cut and pare for three years, meaning one must learn and practice for a long time before being able to produce strips that are consistent in thickness and width.

---

133 **提籃**

Hand basket　40×33×33 cm

竹籃編製的優劣取決於是否能善用竹材的張力以及韌性，特別是要彎製此類提籃的手把，更需要熟練的技藝。

The quality of workmanship depends upon the ability of the artisan to utilize the elasticity and toughness of the bamboo. A high degree of technical skill is necessary, particularly when bending the bamboo to make handles, such as pictured here.

---

134 **提籃**

Hand basket　44×43×42 cm

常做撿螺仔用，所以也叫螺仔籃。

Since these baskets were used for collecting snails, they were also called snail baskets.

1
3
1
◀

1
3
2
◀

1
3
3
◀

1
3
4
◀

134-1 此為圖116成對的菜擔特寫，過去農家菜田所種的蔬果，除了供給自家人吃以外，多餘的則都會用這類菜擔挑到鄰里附近販賣。

In the past, farmers grew vegetables in part to feed their own family. Most of the vegetables, however, were taken to sell in carriers such as this one.

134-2 本省的菜籃形式有很多種，一般皆為平底、圓口，深淺不一，提把則有的用籐做成，但大多仍是用竹條製成。本圖的菜籃形式較常見於中部地區，詳細尺寸規格如圖136。

Many kinds of vegetable baskets are found in Taiwan. Most have flat bottoms and round mouths. Their depths are usually different. Although some handles were made of rattan, most used bamboo. The style of the baskets pictured here is mostly found in the central region of the island. Details and measurements are similar to fig.136.

135 **提籃**
Hand basket 60×44×44 cm
籐把，多做挑菜用，也常用來裝提清洗的衣服。
This basket, with a rattan handle, was used mainly for carrying vegetables. It could also be used as a laundry basket.

136 **菜籃**
Vegetable basket 68×48×48 cm
籐把菜籃，挑菜用。
This basket, with a rattan handle, was used to carry vegetables.

137 **提籃**
Hand basket 70×46×46 cm
此件籃身較大，當不用時，掛於室內可做置物籃用。
This basket has a larger bod. When not used to carry items, it could be hung inside the house and used to store items.

138 **提籃**
Hand basket 50×33×33 cm
也是菜籃的一種，竹製提把俐落的線條，充滿美感。
This is another type of vegetable basket. The bamboo handle is very well made and exhibits beautiful lines.

114

謝 籃 篇

謝籃又叫做「盛籃」，以本省話來發音，二者非常接近，是眾多竹編禮籃中的一種，依本省民間習俗，凡結婚喜慶、赴廟祭拜謝神、中元普渡或向長輩祝壽等，都必須用謝籃來裝置禮品或祭品，以示吉祥或表敬意，所以傳統的竹編謝籃要比大多數的日常竹編器具來得華麗精巧，尤其注重裝飾之鮮麗、上色點漆、文字彩繪等，製作時全得依靠手工及嫻熟的技巧，選用堅嫩適宜的桂竹或麻竹，將其劈成細竹條，按架構逐層穿繞而成，再裝上握柄及配編蓋子，最後在把手彎曲或易折損之處巧妙的打上一個竹結以為保護，就算大功告成。編工之精緻，幾乎讓人找不到任何一絲的竹條接頭。

竹編的禮籃中另有媒人籃、檳榔籃、炮籃、層籃等不同的區別：
媒人籃是婚嫁禮儀中，媒人手提的一種小型禮籃，一般以桂竹精編而成，並上漆或墨繪花鳥圖案，或寫上「百年好合」等討喜字句。

檳榔籃是所有竹編禮籃中最為小巧者，用於喜慶場合中裝盛待客佳品——檳榔，以討相敬如賓之美意。

炮籃是廟會喜慶中用來擺放金銀紙、香燭和禮炮的，體型較檳榔籃及媒人籃都大，提炮籃者一般都走在隊伍之前，沿路點放鞭炮，做開路引導。

層籃是所有禮籃中體型最大的，有二層、三層、四層等不同，每層都由細竹條編成，並留有小孔洞，既美觀且通風，將盛放之物分層放置，可免被壓損，籃柄是用精挑的大竹片做成，從底層連繫中層到上層，再在籃柄中央以鐵片或籐篾編成環鉤，扁擔穿過環鉤即可挑起，從前婚禮中的千金大餅或聘禮就是雇人用層籃挑送的。

今日時代改變了，傳統禮俗日益簡化，竹編禮籃的手藝也已面臨絕跡，人們雖然延續著喜慶廟會，但是手中的竹編禮籃卻早已悄悄地改成大量翻模的塑膠製品。科技雖然進步，機器卻無法代替傳統的竹編手藝，老一輩的匠師已漸凋零，年輕一代怕苦又接不了棒，美好的竹編工藝看來只有讓它失傳了。

Ceremonial Baskets

Xie lan ceremonial baskets were used to hold gifts or offerings for important occasions such as weddings, temple processions, ghost month ceremonies, or the celebration of an elder's birthday. As symbols of fortune or respect, they were more refined and ornately decorated than baskets for daily use. They were usually painted, lacquered or decorated with Chinese characters. Hand-crafted with expertise, they were made from sturdy, young guei zhu and ma zhu, pared into slender strips and woven around a frame. Handles and covers were added to the body and the final touch was to reinforce the handles and weaker spots. The baskets were woven with such skill that it is difficult to spot the ends of the bamboo strips with the naked eye.

There were also matchmakers' baskets, betel nut baskets, fireworks baskets and tiered baskets. The matchmaker's basket was a small basket carried by the matchmaker during a wedding ceremony. Most were woven with guei zhu and lacquered, painted with bird and flower designs or decorated with the characters "bai nian hao he," a wish for a harmonious, long-lasting union.

Betel nut baskets were used during the wedding celebration to treat guests to betel nuts. This tradition also expressed an auspicious wish for the newlyweds - that they respect each other as they would guests.

Fireworks baskets were loaded with gold and silver paper money, incense burners and fireworks for temple festivities. They tended to be larger than the previous two types of baskets. The bearer of the fireworks basket led the procession to the temple, lighting firecrackers along the way.

Tiered baskets were the largest of the ceremonial baskets. The two-, three- and four-tiered baskets were woven of fine strips of bamboo and designed with holes so that air could circulate through them. Items placed in the baskets could be divided among the different tiers to prevent them from being crushed. The handle, made of a large piece of bamboo, extended from the bottom tier to the top, connecting the tiers in between. An iron or rattan ring was attached to the top of the handle so the basket could be carried on a pole. Wedding cakes and betrothal gifts were sent to families using this type of basket.

Traditional customs have now become simplified and the art of bamboo basket weaving is facing extinction. Celebrations and temple festivals have not died out, but the use of hand-crafted bamboo ceremonial baskets has slowly given way to easily available plastic products. Although we now enjoy a state of advanced technology, machine-made products can not compare with traditional bamboo handicrafts. The older generation of artisans is gradually passing away and the younger generation is unwilling to take up the trade. It seems as though the tradition of making these bamboo handicrafts is destined to be lost.

139 層籃

Tiered basket　60×40×40 cm

這一件是清末廣東師傅所編製的層籃，早年隨移民群來台，在九二一大地震後，從東勢災區裡整理出來的，原本應為一對雙層的謝籃，可惜僅剩此件且缺了一層。

This basket was made by a Guangdong artisan during the late Ching dynasty and was brought over to Taiwan by early immigrants. After the 921 Earthquake, it was recovered from the disaster area. It was originally one of a pair of two-tiered baskets, but now only this one remains and it is missing a tier.

圖139層籃的提把兩側如右圖，清楚書著「光緒拾玖年（1893）歲次癸巳□月日」及「許石生置」。

The two sides of the handle of fig. 139 are clearly inscribed with a year corresponding to 1893 (the nineteenth year of the reign of Guangxu) and the name Xu, Shi-sheng.

140 層籃一對

Pair of tiered baskets　54×41×38 cm

以細竹篾精編成上下兩層，東西分層擺放可免被壓損，上層編有小孔，一來美觀、二來通風。

These baskets were finely woven using slender bamboo strips. Separating items into the two tiers of the basket prevented them from being crushed. The small holes in the top tier are both for decoration and ventilation.

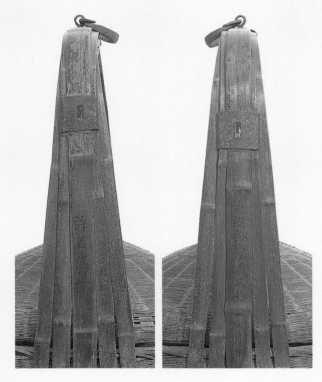

139 · 1▲   139 · 2▲

141 層籃一對

Pair of tiered baskets 75×48×46 cm

此種層籃在舊式婚禮中，以挑送禮餅居多，現行的婚禮雖已無人採用，但其玲瓏有緻、古意盎然的外形，卻在設計界一片復古風中，成為新寵。

This type of tiered basket was used to send wedding cakes to family and friends. Although now no longer used, these exquisitely refined baskets are now popular among designers with a penchant for antiques.

圖141的分合特寫，這對層籃各編有三層，層與層間接合緊密，提柄上釘有鉤環，可以手提，也可以用扁擔挑。

This pair of baskets (fig. 141) have three tiers. Each tier fits perfectly into the other. These baskets could be carried by hand or attached to a pole by the ring on the top of the handle.

圖141的夾層底部特寫。

Detail of fig. 141, bottom of a basket tier.

142 層籃一對

Pair of tiered baskets　86×54×50 cm

這對層籃的做工極為講究，籃身以桂竹劈成的細竹條精編，提柄則以大竹片及藤皮編製，籃蓋及各層的上下緣還以紅褐色漆裝飾，並貼金箔為點綴，呈現出一股脫俗的大戶貴氣。

The workmanship exhibited in this pair of baskets is extremely skilled. The body of each basket is made of guei zhu cut into slender strips and finely woven together. The handles are woven strips of bamboo and rattan skin. The lid and borders of each tier are decorated with reddish-brown lacquer and accented with gold foil. This basket embodies the lofty tastes of the gentry.

右為圖142層籃提把兩側的特寫，上面詳述著「昭和四年（1929）仲秋」及「大內楊協福置」，說明這對層籃原屬台南縣大內鄉的望族楊家所有。

Views of the two sides of the handle of fig. 142. The inscriptions read "Mid-autumn, fourth year of Showa (1929)" and "daneii Yang, Xie-fu," revealing that the baskets were the property of the Yang family of a aristocratic clan in Tainan county.

右為圖142層籃以扁擔挑起的特寫。

Tiered baskets of fig. 142 with carrying pole.

142・1▲   142・2▲

144 **禮籃一對**

Pair of ceremonial baskets　68×38×38 cm

禮籃在喜慶祭祀時可供作挑運禮餅、牲禮及豬腳麵線用，平時還可用來挑送飯送菜點心給田間工作的人吃。

This type of basket was used to carry cakes, sacrifices or pig's feet noodles for occasions of celebration or worship. They could also be used to carry food to workers in the fields.

143 **禮籃一對**

Pair of ceremonial baskets　80×57×54 cm

這對禮籃做工精細，籃身一面繪有花鳥、石榴等吉祥圖飾，一面寫著「福祿壽」、「富貴春」及「囍」等字樣，並題有「丙辰年（1916）郭嬰」是從前餅行出借給新人挑運禮餅專用的。

This pair of baskets is exquisitely crafted. Auspicious paintings on one side of the body include birds, flowers and pomegranates. The other side is inscribed with the blessings of longevity, prosperity and happiness, as well as the year corresponding to 1916 and the name Guo Ying. The baskets were used to send wedding cakes to friends and family.

145 謝籃

Xie lan ceremonial basket 35×32×30 cm

此類謝籃原作炮籃使用，多數時候也做爲赴廟進香裝
金燭用。

This type of basket was originally used for carrying
firecrackers, but was also used to carry candles in temple
processions.

146 謝籃

Xie lan ceremonial basket 37×37×33 cm

此類髹黑、紅色漆的謝籃，專爲本省客籍地區所用，
形式大方，獨具美感。

This type of black and red lacquered basket was used in
the Hakka districts of the island. The form is generous and
unfussy with distinctive beauty.

147 謝籃

Xie lan ceremonial basket 36×36×36 cm

作炮籃用，ㄇ字形的提把彎轉處，特繫有竹結，以減
緩折裂受損。

The strips which make up the arched handle on this
firecracker basket are tied together with bamboo. This
added touch strengthens the handle by making it less
prone to snapping where the bamboo has been bent to
form the curve.

145 ◀

146 ◀

147 ◀

148　**謝籃**

Xie lan ceremonial basket　36×21×20 cm

作檳榔籃用，供裝檳榔、香煙招待來賓。

This basket was used to offer betel nut and cigarettes to guests.

149　**謝籃**

Xie lan ceremonial baskets　28×18×17 cm

另一種形式的檳榔籃。

This is another style of betel nut basket.

150　**謝籃**

Xie lan basket　54×40×40 cm

祭祖拜神時供裝三牲、四果、麵線等祭品用。

This basket was used to hold offerings for worshiping ancestors and deities, such as the three sacrificial offerings (ox, sheep and hog), the four fruits or noodles.

151 **謝籃**

Xie lan ceremonial basket　42×30×26 cm

作媒人籃用，在結婚場合中供媒人提用，籃蓋上繪有
八卦紋，可驅邪避凶。

This matchmaker's basket was carried at a wedding. The
lid is painted with a design consisting of the Eight
Diagrams to ward off evil.

152 **謝籃**

Xie lan ceremonial baskets

（1）46×38×38 cm　（2）10×9×9 cm

（1）由桂竹精編的謝籃。

（2）婚禮陪嫁用的迷你謝籃。

(1) This basket is beautifully crafted of guei zhu.

(2) This tiny basket was used for the marriage dowry.

圖152（1）提把彎轉處的竹結特寫。

Detail of the bamboo knot on the curve of the handle of
fig.152 (1).

(1)

(2)

禮 器 篇

在台灣住民的宗教禮俗中，無論是祭拜神明、祖先或歲時祭典，都會用到許多富含民俗藝術價值的竹製用品，最普遍易見的是一對半月型的占卜用具，閩南話叫做「杯」或「筊」，是以竹子的地下莖剖半製成，用來當做人神之間的溝通媒介。在台灣，凡有廟就有筊，香火愈盛的廟，筊也愈多，一般信佛的家庭也都會備有筊，在祭拜時以擲筊來請示神明的意旨，如果一對筊出現正反兩面，就表示神明答應或認可，若是都成平切面，即是笑杯，表示模稜兩可，或無可奉告；反之，若是出現兩凸面，則為怒杯，表示不可或預示凶機。

另一個常見的竹製禮器是籤筒及籤支，籤支是竹片削成，籤筒是以麻竹或孟宗竹的竹稈製成，透過擲筊取得神明的認可後，即可從籤筒中抽出一支籤支，再對照籤詩，就能更明白神明的意旨了。一直以來，台灣人透過擲筊及抽籤的占卜，做成了無數的重大決定，影響台灣人的命運至為巨大。

今天除了竹製的香筒及香腳仍是大多數人常用的以外，其他的竹製禮器都不易找到了，像是利用竹篾編成的各式供盤、水果盤、花籃等，多已被塑膠品取代；而特殊祭典才用到的牲禮架、鼓架及祭祀時專盛棗栗桃梅之屬的籩豆等，則更是幾近絕跡。

Ritual Objects

No matter what the occasion for worship, the Taiwanese used a variety of bamboo ritual objects that are also valuable examples of folk art. Divination crescents fashioned from bamboo roots were the most commonly seen type of ritual object. These divining tools can be encountered in every temple, and the livelier the temple, the more there are. Buddhist households also often have a set. The crescents act as a medium between the human and spiritual realms, and by consulting them, one can receive advice from the gods.

The crescents are interpreted from the way the pair land after they are tossed. Each crescent has a flat side and a rounded side. If they fall with one flat and one rounded side face up, the answer is affirmative or positive. Both flat sides facing up reveals the answer to be ambiguous, while the appearance of two rounded sides represents a negative response.

Bamboo chips and containers to hold them are other common bamboo ritual objects. The containers were made of stalks of ma zhu or mengzong zhu. After using the bamboo crescents, a chip is selected from the container. Each chip bears an inscription which further clarifies the meaning of the divination session using the bamboo crescents. Even today, this method of divination has a great impact on a person's fate and is used when making important decisions.

Aside from objects for incense, other bamboo ritual objects are no longer easy to find. Woven bamboo offering plates, fruit plates and flower baskets, for example, have all been substituted with plastic ones and ritual items used only for special ceremonies such as sacrificial stands, drum stands and vessels for dates, chestnuts, peaches and plums have long disappeared.

153 籤筒

Divination slips and container　40×11×11 cm

本省民衆過去有求於神明時，總是到廟裡燒香膜拜，卜卦抽籤，由籤詩來論吉凶，一般常抽的運籤如本圖，按天干、地支組合共六十隻，外加籤王合計六十一隻。

When seeking advice from the gods, people would go to the temple to burn incense, kneel in worship, perform divinations and pick divination slips. The inscriptions on the slips would help to predict one's fortune. The slips used were like those pictured here. Sixty-one slips were customary, in accordance with the Ten Celestial Stems and Twelve Terrestrial Branches, which form a cycle of 60, plus one.

154 籤筒

Divination slips and container　16×11×10 cm

在較大型的寺廟除設有運籤外，另設有藥籤供信衆求取藥方，藥籤筒一般較運籤筒來的小，籤支以數目編號，並像醫院般分有大人藥籤、小孩藥籤及眼科或婦科等類別。

In addition to fortune-telling slips, large temples also offered medical slips which contained prescriptions. The containers for these slips were usually smaller than those for fortune-telling slips. Each medical slip was numbered and was divided into prescriptions for adults and children, and for categories such as ophthalmology and gynecology.

155 香筒一對

Pair of incense containers　29×15×14 cm

香筒一般都以麻竹或孟宗竹粗大的竹稈製成，成對的香筒通常爲廟裡所使用，此對刻有哪吒神話題材的香筒，甚爲特殊。

Incense containers were made of thick, large stalks of ma zhu or mengzong zhu. They were found in pairs inside temples. This pair is distinctive for the carving of a traditional folk tale concerning the deity Na Zha.

156 筊杯二對

Two pairs of divination crescents

（1）8×4×2 cm　（2）21×13×4 cm

竹製筊杯是少數未被塑膠取代的禮器，因爲塑膠做的筊杯太過滑手，擲地後難以收拾，不若竹筊來得耐擊、耐磨，所以人們至今仍偏愛竹筊。

Very few divination crescents are made of plastic because they are too slippery and difficult to control. Ones made of bamboo are durable and resistant to wear so even today, bamboo ones are still preferred.

1
5
3
◀

1
5
4
◀

1
5
5
◀

1
5
6
◀

157　**花器**

Flower vessel　30×14×14 cm

供祭祀插花用。

Flowers were put in this vessel and used when worshiping.

158　**籩豆**

Food vessel　21×15×15 cm

祭祀時裝放棗栗等雜糧專用。

This vessel was used to contain offerings such as dates and chestnuts for the purpose of worship.

159　**供盤**

Offering dish　33×33×13 cm

供放牲禮、敬果用。

This type of dish was used for sacrificial offerings and fruit.

160　**祭果盤**

Fruit offering dish　25×25×8 cm

裝放水果用。

Offerings of fruit were placed in this type of dish.

157 ◀

158 ◀

159 ◀

160 ◀

161 **牲禮架**

Sacrificial stand　74×51×41 cm

神明壽誕或家中長輩大壽祭祖時，多以此牲禮架披掛
長條麵線，代替殺牲，祈求長壽。

When celebrating the birthday of gods or elders, these
stands were strung with long strands of noodles as a
symbol of a sacrificial offering and conveyed the blessing
of longevity.

圖161側面特寫，造形充滿美感，猶如雕塑品一般。

Side view of fig. 161. The stand has a sculptural beauty.

162 **牲禮架**

Sacrificial stand　124×70×58 cm

又名豬公架，是特定廟會賽豬公祭神時，用來撐開整
隻屠宰後的豬隻所用。

Also called a pig stand, this was used to prop open a
slaughtered pig as a sacrificial offering for special temple
festivals.

163 **鼓架**

Drum stand　115×42×38 cm

這是迎神賽會所使用的鼓架，三人一組以扁擔挑起鼓架，其中一人敲鑼，一人打鼓，另一人擊鈸，走在遊行隊伍前爲神明引路。

These stands were used for processions and festivals in honor of the gods. They were carried by a group of three people who led the procession. One person would strike a gong, one would beat the drum and the third would play cymbals.

圖163鼓架及所裝置的鼓、鑼及鈸一組。

The drum stand of fig. 163 and drum, gong and cymbals.

164 **鼓架**

Drum stand　63×47×31 cm

這型鼓架以細竹管編製而成，是屬戲團樂隊所用。

This type of drum stand, made of slender, tubular sections of bamboo, was used by musicians in theatrical troupes.

畫筒

Painting container　97×24×24 cm

這是民間神壇法師用來裝神像畫軸用的畫筒，通體以細竹篾編成，再以粗竹條箍住，四肩釘有小鐵扣，法師外出做法會時，即穿上繩子以扁擔挑運。像這類的畫筒一般並不多見，普通神壇所用的畫筒都是用鐵片製成的，而從這件竹製畫筒外觀所題的書法看來，這畫筒原來應屬於一個頗為講究的神壇。

These containers were used by priests at small temples for portraits of gods. When the priest left the temple for certain ceremonies, these containers could be transported with him on carrying poles. The container here is woven with fine strips of bamboo, bound with thick strips and affixed with four small iron clasps. This type of container is not very common. They are usually made of pieces of iron instead. Furthermore, the calligraphy on the sides of the container shows it's a testament to the erudition of the original owner.

搖籃篇

竹子是上天賜給台灣人民的恩物，竹類資源從南到北遍佈全台，由於竹子具有生長快速的特性，栽種三到五年即可被利用，加上取材容易，質輕而堅，早在台灣一般的家庭中被普遍製作成各式各樣的器具，無論是枝、葉、竿、簀、篾等全都可供作精緻的竹藝品，與早期台灣居民的一生可以說有著密不可分的關係，像是嬰兒初生時所睡的竹搖籃，即是以竹篾所編成，這類的竹搖籃形似橢圓盆狀，底部略平，使用時只須在裡面鋪以小被褥，就可成為嬰兒舒適安穩的睡床，過去農村的「台灣之子」幾乎都是睡竹搖籃長大的，只是在全省各地因地域的分隔，所編出來的搖籃在大小高矮等規模稍有差異而已，另在風大日烈的南部地區所編的竹搖籃常會帶有頭蓋，可預防風吹日曬。

竹搖籃安置的方式，最常見的是以粗繩索或舊腳踏車內胎直接繫於屋樑上，或另以桂竹做成折合式搖籃架，將搖籃繫在搖籃架上，左右輕輕擺動，很快就能讓幼兒進入夢鄉，農忙時節，還可連同搖籃架一併搬到田邊的陰涼處，就近一邊照顧小孩，一邊工作。

等到幼兒稍長，開始會爬學坐時，「椅轎」便是最佳良伴，這種專為幼兒設計的竹椅，四周可將幼兒圍起，使其不會倒下，另一種設計是大人小孩都適用的「乳母椅」，此種椅子側放時即成轎狀，小孩入座十分平穩，座前並設計有竹圈圈可供玩樂，及鋪放的竹板可擺放碗筷和玩具，若是幼兒突然有便溺時，也只要將小孩抱起，用水將竹椅沖淨擦乾就好，非常堅固耐用。

「乳母車」是以桂竹稈榫接成骨架，再將竹稈敲成平板鋪做床面，配上四輪的嬰兒車，依需求不同可做坐、臥兩種調整，在過去也是家庭照顧嬰幼兒必備的用具。最後，在沒有紙尿布及烘衣機的那個年代，多雨濕潤的北台灣還得靠竹篾編成的烘尿布籠來烘乾尿布。這些天然的竹製嬰幼兒用品不知伴隨了多少人一路成長到大，這種幸福大概是二十一世紀的新生兒所體會不到的吧！

Cradles

Taiwan is blessed with an abundance of bamboo. A fast-growing plant, which can be used after only three to five years of growth, easily harvested and lightweight yet durable, bamboo was used to make all sorts of household objects early on in Taiwanese history. Every part of the bamboo plant could be used -branches, stalks, leaves, roots and strips. Bamboo was inseparable from the lives of the Taiwanese in early times and accompanied them from the cradle to the grave.

Cradles were woven from bamboo strips and were shaped like oval basins. The bottom of the cradle was flat and when covered with a blanket, made a comfortable and secure bed for the baby. Sleeping in a cradle such as this was a common experience for nearly every child in the rural villages of the past. Although the basic form of the cradle was fairly uniform throughout the island, there were slight regional differences, such as variations in height and size. In the southern districts of the island, where strong winds and harsh sun are common, cradles were designed with hoods to shield the child.

The cradle could be tied to an interior beam using a thick rope or rubber tires from old bicycles, or suspended from a folding stand made of guei zhu. Thus suspended, the cradle could swing back and forth, lulling the baby to sleep. During the busy farming seasons, when everyone went out to the fields, the cradle and folding stand could be brought out and set up under a shady tree. This way people could keep an eye on their children while working in the fields.

When children began to crawl and sit up, they graduated to a baby chair. These chairs enclosed the child on all four sides so it couldn't fall out. A similar type of chair, called a "mother and child chair," could be used by both adults and small children. When turned on its side, the chair became a baby chair like the type described above. The front of the seat even had bamboo rings for the child to play with and a table-top to hold the child's food or toys. If the child soiled itself, the chair could be easily cleaned with water and wiped dry.

Baby carriages were other essential child-rearing devices made of bamboo. Using guei zhu stalks and a tenon and mortise system, a frame was constructed upon which flattened slats of bamboo stalks were laid to make the bed. The carriages were adjustable so that the baby could either sit or sleep. Finally, bamboo diaper drying baskets were a necessity in a time before disposable diapers and dryers existed, especially in rainy, humid northern Taiwan.

Countless Taiwanese were reared with these natural bamboo products, however, this is an experience children of the 21st century will never have.

166 搖籃

Cradle　88×46×38 cm

搖籃普遍以細竹篾編六角孔做成，形似橢圓盆狀，過去多數由婦人自己編做，但也有竹細工店販賣現成的，樣式間略有差異。

Cradles were customarily shaped like oval basins and woven with thin bamboo strips in a pattern leaving hexagonal holes. There are, however, many variations on this model. In the past, many women would make their own cradles as well as buy those made from workshops.

167 搖籃

Cradle　80×40×40 cm

綁上繩索的搖籃，可任意掛置，忙碌時掛在屋簷樑下或田邊樹下都能方便就近照顧。

These cradles could be suspended by rope from beams within the house or in the fields under a tree. It was thus possible to keep the child close at hand even when the parents were busy.

168 搖籃及搖籃架

Cradle and Cradle stand 　（Cradle）84×56×22 cm
（Cradle stand）116×110×9 cm

當無處可吊掛搖籃時，如在空曠的庭院，搖籃架便是最佳的搭擋。

Stands were used when there was no other place to hang the cradle.

169　烘尿布籠

Diaper-drying basket　70×43×32 cm

將竹籠覆蓋，內置一火爐即可用來烘尿片或一般的衣物。

These baskets were shaped like an overturned basket or lid. A heat source was placed inside to dry diapers or clothing, which were placed on top.

170　搖籃

Cradle　75×49×14 cm

這件搖籃的邊圍特別低，竹篾也比較細小，較適合小女嬰所用。

This cradle with particularly low sides is made with thinner strips of bamboo, making it more suitable for infant girls.

171　搖籃

Cradle　103×50×27 cm

這件長度超過一百公分長的搖籃，是本書收錄的十件搖籃中，最長的一件，是較為少見的。

The length of this cradle exceeds 100 cm and it is the longest of the ten cradles recorded in this book. It is a rarer example.

172　搖籃

Cradle　100×57×24 cm

這件搖籃編製得較為綿密耐用，可以讓大家庭中姑嫂妯娌等多人輪流共用。

This cradle was woven with great attention to detail and durability. It could have been used in turn by all of the women of a large household.

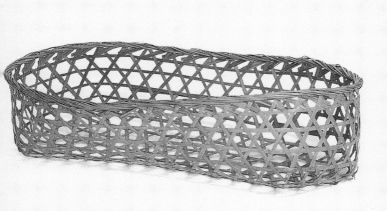

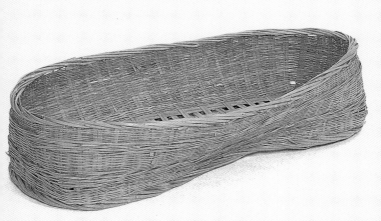

173 **搖籃**

Cradle　81×54×20 cm

搖籃的製作，有時是出自初爲人父的男人之手，在有樣看樣，無樣自己想的情況下，所編出來的形式、大小也就各有不同。

There are many variations in form and size among cradles. Fathers who handcrafted them for their own children either modeled the cradles after existing styles or made them according to their own specifications.

---

174 **搖籃**

Cradle　74×51×18 cm

這件搖籃的竹皮色澤深厚發亮，想必是有較長的使用年代。

The deep hue and glow of the bamboo surface indicates this cradle had a long history of use.

---

175 **搖籃**

Cradle　76×65×24 cm

這件搖籃比一般的寬闊，有六十五公分寬，按推想，可能是雙胞胎用的搖籃。

The proportions of this cradle are more generous than most with a width of 65 cm. Perhaps it was used for twins.

---

176 **搖籃**

Cradle　76×52×38 cm

這件搖籃邊圍較高，有防嬰孩受風著涼的功用。

The sides of this cradle are particularly high to shield the child from drafts.

1
7
3
◀

1
7
4
◀

1
7
5
◀

1
7
6
◀

177 椅轎

Baby chair 35×27×25 cm

此件椅轎形式優美，年代甚早，色澤亮麗華潤，四面可坐，是早期台灣住民生活竹器中難得的佳作。

This beautifully designed baby chair is of an early period. Years of use have smoothed the surface and given it a glowing hue. All four sides are the same and the child could be seated facing any direction. It is hard to come across such superb early pieces of Taiwanese furniture.

---

178 乳母椅

"Mother and child" chair 47×41×41 cm

乳母椅設計巧妙，大人小孩都適用，圖為側放之乳母椅，適合周歲左右幼童乘坐，非常平穩安全。

The ingenious design of this chair made it convenient for both children and adults. On its side, as pictured, it serves as a seat for a child of about one year.

圖為直立之乳母椅，供大人乘坐。

Upright, as pictured here, the seat can be used by adults.

---

179 乳母車

Baby carriage 96×80×53 cm

此種乳母車可調整成坐、臥兩式，並有四個輪子可推移，居家外出都適用。

This type of carriage could be adjusted so that the child could either sit up or lie down in it. It was useful when go out with the child.

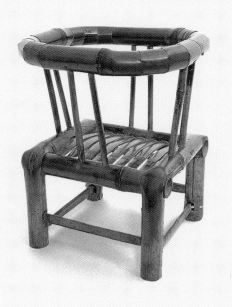

177 ◀

178 ◀

178.1 ◀

179 ◀

食 器 篇

在傳統的竹編器具中，日常飲食及炊煮用具也爲數不少，像是三餐最不可缺的竹筷、裝箸用的筷子籠、飯後剔牙用的牙籤、吃碗粿用的籤仔以及盛飯用的飯匙等等不勝枚舉，由於竹材輕便、耐用、且可塑成的造形樣多，因此除了能滿足實用的功能外，多數的食器造形也能間接傳達出先民樸實的情感及他們心目中美觀的形象。

台灣人向來節儉，不喜奢華，尤其昔日農村社會裡三餐多半吃著粗茶淡飯，只有逢年過節以及客人來訪時才在大廳中以雞鴨魚肉招待，平時廚房往往就兼具了飯廳的功用，廚房內的擺設一般有竹菜櫥或飯桌（兼菜櫥）、桌罩、碗籃、蒸籠及其他各式功用的竹籃。

竹製的菜櫥以竹管編鑿而成，依高度不同有二層、三層等分別，還有一種約三尺高的飯桌式菜櫥，即可當飯桌用也可存放菜餚，具有雙重功用。

竹編桌罩呈盒蓋形，尺寸比桌面略小一些，桌罩的側邊以細竹篾編成細格子狀，能利於通風，減緩食物腐壞，正上方則以較大的竹篾密編成蓋，可防止從上而落的灰塵或蟲蠅沾染食物，在紗窗還不普及的年代，竹編桌罩是家庭的必備品。

將洗乾淨的碗盤放在竹子編成的碗籃裡可自然風乾，兼具環保概念及通風的功能，在塑膠製品出現以後，現已近絕跡了。

竹製蒸籠是家家必備的廚房炊具，逢年喜慶都會用它來製作糕點以供拜神明，一般分有方形和圓形兩種，編製的竹材是桂竹，製作的過程相當繁複，從選竹、劈竹、削竹、削片、鑽孔、綁籐、磨平到編結，都須良好的技術與經驗，否則不堪多久的使用，就會變型毀壞，現今仍能製作蒸籠的老師傅已所剩無幾，而電器化的烤箱和微波爐卻早已取代了蒸籠的地位。

在冰箱尚未普遍使用之前，廚房樑上都會吊個幾只吊籃用來擺放剩菜或糕餅等，不僅小孩拿不到，連貪嘴的貓鼠輩也無法偷吃，所以這類吊籃又叫「氣死貓」，如果再在吊勾上塗上牛油，那就連螞蟻也莫可奈何了，通常這類的吊籃體形較大，略呈半圓球形，有圓口及口蓋，農婦會將做好的飯菜或點心置於吊籃內，以扁擔挑送至田裡給終日在外頭辛苦耕作的男人吃，所以這類的吊籃可說是居家外出皆宜使用。

154

Food Implements

Bamboo implements for food and cooking made up a large portion of the bamboo handicrafts produced. Using bamboo, a convenient, durable and versatile material, to make indispensable objects such as chopsticks, chopstick containers, toothpicks, forks, and rice scoops not only satisfied practical needs but also expressed the Taiwanese sense of simplicity and aesthetics.

A thrifty people who frowned upon extravagance, rural villagers ate simple meals, using the kitchen as a dining room. The kitchen was usually stocked with bamboo food cupboards or tables which doubled as cupboards, table covers, baskets for bowls, steamers and an assortment of other baskets. Only on holidays or when entertaining guests, would villagers eat more luxurious meals of chicken, duck or fish out in a specially-designated dining area.

Two- and three-level food cupboards were made with tubular sections of bamboo. Another type of cupboard was about a meter high and also functioned as a table. Table covers were made slightly smaller than the table top.

The sides of the cover were woven in a fine checked pattern to allow for air to circulate and keep the food fresh. The top of the cover was tightly woven with larger bamboo strips to prevent dust and bugs from falling into the food. In an age when window screens were not widespread, bamboo table covers were a necessity.

After the dishes were washed, they were placed in bamboo baskets to dry naturally. This environmentally-friendly device disappeared from use without a trace with the introduction of plastic products. Bamboo steamers were another staple in every household. On festivals, they were used to make cakes as offerings for the gods.

Steamers were made from guei zhu and came in two shapes - round and square. Producing them was a complicated process in which bamboo was carefully selected, chopped up, pared, cut into pieces, perforated, bound, smoothed and then woven. The process required great skill and experience or else the finished product would end up warping and falling apart after only a few uses. Very few artisans with the skill to make these steamers exist today. At any rate, ovens and microwave ovens have made them virtually obsolete.

Prior to the use of refrigerators, people stored leftover food and cakes in baskets hung from kitchen beams. This not only prevented children from getting into the food, but also cats and household vermin, thus the basket was called "infuriate the cat." By spreading a little butter on the hook holding the basket, even ants couldn't get into the food. These types of baskets tended to be larger than other food baskets and were half-spherical with a round mouth and cover. Women also attached these baskets to a carrying pole to bring food out to the men working in the fields.

180 （1）箸筒 （2）箸筒底部
（1）Chopstick container (2) Bottom of container

21×8×8 cm

此類箸筒是過去民家擺放筷子常用的，以竹筒製成，
上端留有穿孔的吊把可供吊掛，底部鑿洞有便於風乾
洗後的竹筷。

Earlier Taiwanese households used these bamboo
containers for chopsticks. It could be hung by the handle
with a hole at the top end. The bottom of the container
with holes for air to circulate. After they had been washed,
chopsticks could be placed inside the container to dry.

---

181 （1）麵撈 （2）鼎刷
（1） Noodle ladle （2） Pot scrubber

（1）60×9×9 cm （2）17×10×2 cm

麵撈是傳統麵攤煮切仔麵所用的竹笊，現在已少有人
用，幾乎都改用鋁製的了，而竹製的鼎刷絕跡得更早，
全被各種菜瓜布、鐵刷子給取代了。

Nowadays, ladles such as this are made of aluminum
instead of bamboo. Bamboo pot scrubbers disappeared
even earlier as sponges and metallic brushes became
common.

---

182 飯濾
Strainer 55×31×16 cm

以前人生活困苦，主食都以甘藷簽拌白米，而飯濾就
是炊煮時淘取的用具。

In earlier times, people led lives full of hardship. Meals
consisted mainly of rice cooked with sweet potatoes and
were ladled out with this utensil.

---

183 （1）油飯叉 （2）飯匙 （3）飯匙 （4）碗粿簽仔
（1） Rice fork （2） Rice spoon （3） Rice spoon
（4） Glutinous rice pick

（1）17×6×3 cm （2）20×7×0.5cm
（3）31×6×0.5 cm （4）16×1.5×0.2 cm

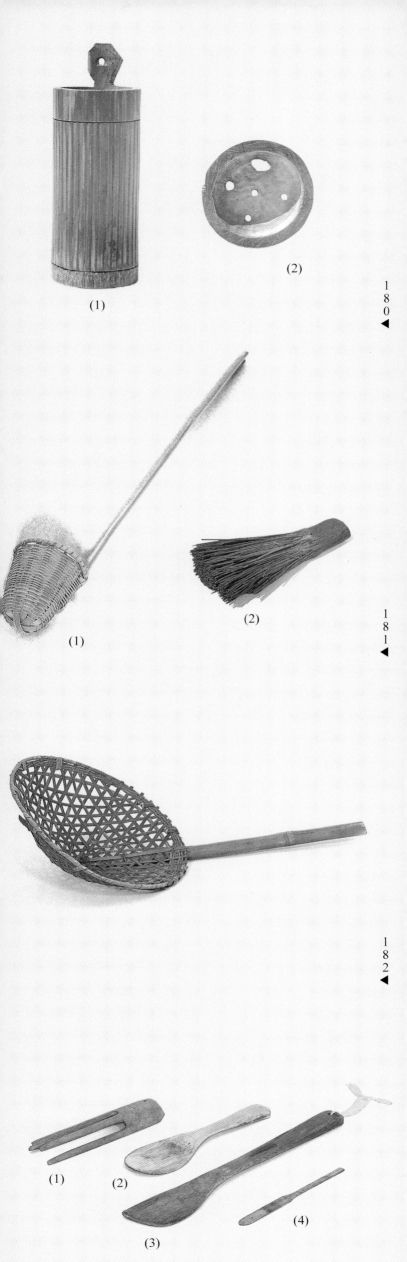

(1)

(2)

180 ◀

(1)

(2)

181 ◀

182 ◀

(1)   (2)

(3)

(4)

183 ◀

184　（１）蒸片　（２）透氣筒

（１）Steamer mat　（２）Ventilating tubes

（１）56×56×1.5 cm　（２）13×4×4 cm (6)

在蒸煮時，蒸片是用來墊開食物與沸水的，透氣筒則
專用於密閉的蒸籠中，使熱氣得以進入蒸熟。

Steamer mats were used to separate the food from the
boiling water used to steam them. The tubes were used in
sealed steamers to direct the hot air into the steamer.

185　**蒸籃**

Steamer basket　54×52×13 cm

這是一般四方蒸籠內用來裝粿的方籃。

This basket was placed inside of rectangular steamers for
steaming glutinous rice.

186　**蒸籃**

Steamer basket　60×60×15 cm

可用於蒸煮菜餡或糕粿。

This type of basket was used to steam vegetables or
glutinous rice cakes.

187　**醬油桶**

Soy sauce tub　41×41×38 cm

這是半世紀以前興用的醬油桶，桶身以木片一一圍攏
後，再從外以竹條箍緊，這樣就能止滲不漏了。

These containers were used in the first half of the century.
The bodies were made of wooden slats bound together
with bamboo which prevented leakage.

158

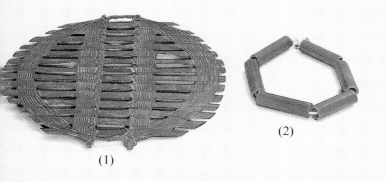

(1)

(2)

188 **大吊籃**

Large hanging basket　60×54×54 cm

俗稱「氣死貓」，懸掛於樑上，存放食物用。

These were also called "infuriate the cat." Food was stored inside the baskets which hung from beams inside the house.

189 **大吊籃**

Large hanging basket　50×50×50 cm

與上圖113同屬本省客家式的「氣死貓」。

This style of hanging basket was used by the Hakka in Taiwan.

190 **大吊籃**

Large hanging basket　68×62×62 cm

閩南式的「氣死貓」，多為河洛人所用。

This southern Fujian-style basket was mainly used by people from Fujian.

191 **菜剉支架**

Vegetable paring stand　62×51×35 cm

菜剉支架上通常綁有剉刀板，是過去農人用以剉甘藷
簽或蘿蔔絲的器具。

This type of stand was usually equipped with a paring
board and was used to pare sweet potatoes or carrot
shreds.

192 **方形蒸籠**

Square steamer　66×66×40 cm

此類蒸籠是農業社會時，家家必備的廚房炊具，外框
是以檜木製成，裡面的蒸片則以桂竹編製，逢年過節
都會用它來製作各類糕點。

In Taiwanese society many years ago, every household
owned one of these steamers. The outside was made of
kuai mu (Chinese juniper or cypress) and the mat inside of
guei zhu. They were used to steam cakes for New Year's
and other festivals.

193 **圓形蒸籠**

Round steamer　43×43×31 cm

此款蒸籠製作不易，必須要懂得選材、選竹、劈竹、
削竹、鑽孔、磨平、編綁等功夫才行，若是技藝疏陋，
蒸籠很快就會鬆垮變形，不堪使用。

Making round steamers was not easy. The artisan had to
be skilled in choosing the materials and the type of
bamboo, and know how to cut, pare, bore holes, smooth
and weave the bamboo. If the artisan's technique was
lacking in any of these respects, the components of the
steamer would soon loosen and warp, making it useless.

191
◀

192
◀

193
◀

194 **鼎蓋**
Pot cover　72×72×40 cm
傳統大灶專用。

These were traditionally used on large stoves.

---

195 **蒸床**
Steamer bed　72×72×7 cm
糕餅店蒸麵龜專用。

Shops selling cakes and bread used this type of steamer.

---

196 **桌罩**
Table cover　78×78×23 cm
舊社會時代，衛生條件不佳，民屋也少有安裝紗窗，
因此要避免蚊蠅沾染桌上的食物，就得依賴竹編的桌
罩了。

Sanitary conditions were not very good in the past. Most
houses did not have screened windows and this type of
cover was used to prevent flies and other bugs from
contaminating the food on the table.

---

197 **桌罩**
Table cover　77×77×20 cm
上圖196桌罩形式爲中南部所盛行，北部的桌罩形式則
以本圖居多。

The style of table cover in fig. 196 was commonly for
the central and southern regions of the island, while the
north mostly used covers like the one pictured here.

198 儲米籮

Rice storage basket 38×36×36 cm

此類竹籮有蓋，肩上編有四耳，除了能當儲米籮，外出
時在四耳穿上繩索，也可當行李箱用扁擔挑運。

This basket with a lid and four loops can be pick around the
shoulder. Not only storing rice, the basket could also be
carried as luggage by threading a rope through the loops
and attaching it to a pole.

199 儲米籮

Rice storage basket 38×3838 cm

另一種以竹稈做邊，有四腳的儲米籮，因底部與地面有
一固定的距離，可保持通風，避免裡面所裝的物品潮溼
發霉。

This style of the basket has bamboo stalk rims and four
legs to lifting the basket off the ground, in order to
ventilate and prevent the contents of the basket
from molding.

200 碗籃

Dish basket 30×30×18 cm

竹編的碗籃具通風性與滲水性，可使清潔後的餐具可自
然陰乾，蓋上碗籃蓋後並能防止蚊蠅沾染。

This basket is ventilated and allows for the drainage of
water. Washed dishes were placed inside to air dry. The lid
keeps bugs from getting inside.

201 碗籃

Dish basket 31×31×21 cm

形式稍異的另一款碗籃。

This is a slightly different style of dish basket.

202 菜櫥

Food cabinet　82×82×51 cm

此件菜櫥外型俐落大方，條理不紊，裡外共三層結構，
既實用又美觀。

Well-made, naturally elegant and with ordered, clean lines,
this three-tiered cabinet combines practicality and artistry.

右為圖202打開門扇，內裝之特寫。

View of interior of Fig. 202.

203 菜櫥

Food cabinet　85×82×53 cm

這件菜櫥的門扇設計，充分發揮了竹枝的線條特色，同
時也達到櫥櫃通風及由櫥外透視櫥內的功用，是早年竹
製菜櫥中的佳作。

The door panels of this cabinet not only show off the lines
of the bamboo stalks, but also ventilate the cabinet and
allow the contents to be seen without opening the doors.
This is an excellent example of early cabinetry.

204　菜櫥

Food cabinet　184×85×48 cm

這是一件少數保存的相當好的大型菜櫥，非常難得，過去這類菜櫥先是被木製的菜櫥給淘汰掉大半，在冰箱流行後，更是完全被取代，由於體型龐大，室內難容，多數被丟棄屋外任憑腐壞，更有被直接劈來當柴燒的，所以能完好留存至今的，誠屬不易。

There are very few large bamboo cabinets have
been preserved in such good condition. They were replaced
by wooden cabinets firstly, then they were made obsolete by
refrigerators. Since they were so large that took up a lot of
space inside homes, many people discarded them outdoor
spoiled. Some cabinets were also chopped up and used as
firewood.

Therefore, finding such a well-preserved example is
extremely difficult.

205　菜櫥

Food cabinet　81×81×49 cm

這是所有竹製菜櫥中，樣式最簡單的一款，僅有一層櫥櫃，適合居家地方小，成員少的家庭使用。

This is the simplest style of bamboo cabinet. Consisting of
only one level, it was suited for homes with little space
and few family members.

其 他

「竹」為一個象形文字，在中文字典上大部分有「竹」字頭的字都與竹器物有關，而這樣的字有將近兩百個之多，可見竹器物的名目有多繁複，另有一些本省人以閩南語或客家話命名的竹器，是在字典中找不到適當字眼可以表述的，所以民間日常生活所使用的竹器種類之多、範圍之廣，其實是很難估算的，尤其生活形態改變之後，早年的竹器物若沒有得到妥善的保留，到現在根本也就無從瞭解其確切的形貌、功用或名稱，因此，要統合所有的竹器做細目歸類，確實有困難。

本書前面八個篇章是約略以不同的竹器型態或實際功用做分類，而無法納入，或不能自成一類的，僅在此合併收錄於後。

Others

The Chinese character for "bamboo" is a pictograph. Characters which employ the bamboo radical all have something to do with bamboo, and in a Chinese dictionary, there are about 200 words which have the bamboo radical. This fact alone does not even fully indicate of the pervasiveness of bamboo in Taiwanese life, as some Taiwanese and Hakka words regarding bamboo and bamboo objects are not even found in the dictionary. Numerous types of colloquial objects were used for a wide range of purposes, however, the names, appearances or uses of many of these objects have been lost. As Taiwan went through its process of rapid modernization, little effort was made to preserve many of the early handicrafts produced on the island. With such scant information available, cataloguing the objects that have survived is a difficult task. The previous nine chapters based the divisions of the objects on their form or function, however the objects in this chapter could not be classified due to a lack of sufficient information. Therefore, they have been grouped here under the category of "others."

206 竹枕

Bamboo pillow　26×13×10 cm

以竹條製成的枕頭，略帶彈性，睡來清涼舒適，出外旅行也便於攜帶。

The pliant bamboo strips used to weave this pillow make it comfortable and cool. It could also be easily carried for traveling.

207 竹夫人

Bamboo "lady" pillow　93×17×17 cm

竹夫人又叫作抱籠，一般多供男子使用，在酷熱的夏夜中抱著竹夫人睡眠，既祛暑又別富情趣。

Also known as an embracing pillow, this type of pillow was mostly used by men who would embrace the pillow as they slept on hot summer nights to keep cool.

208 （1）竹扇子　（2）洗髮梳

（1）Bamboo fan　（2）Bamboo hairbrush

（1）42×26×0.3 cm　（2）5×5×4 cm

209 （1）筆筒　（2）筆筒　（3）工程尺　（4）台尺

（1）Brush holder　（2）Brush holder

（3）Ruler（4）Taiwan standard measure rule

（1）　12×7×7 cm　（2）12×8×6 cm

（3）31×3×0.4 cm　（4）30×2×0.4 cm

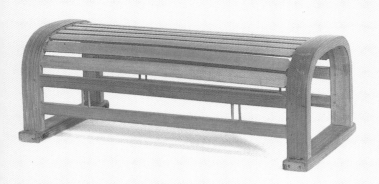

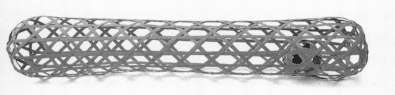

(1)

(2)

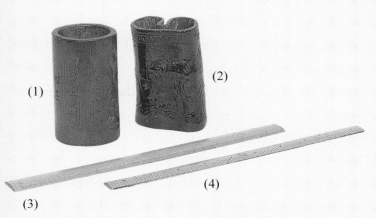

(1)

(2)

(3)

(4)

210　**紙傘**

Paper umbrella　65×48×48 cm

紙傘的傘骨多以彈性佳的孟宗竹削成，傘面則以油紙糊成，在塑膠傘問世以前，紙傘風行全台。

The frame of this umbrella is made of flexible mengzong zhu, the top is made of pasted oil paper. Before plastic and synthetic umbrellas, these were popular over the entire island.

圖210紙傘收合狀，攜帶輕巧。

View of fig. 210 folded in a convenient and easy to carry form.

211　（1）**手提包**　（2）**手提包側面圖**

（1）Bamboo purse　（2）Side view of purse

32×23×10 cm

此款竹製手提包樣式摩登討喜，雖是幾十年前的手藝品，但論其美感完全不亞於現代流行的名牌貨

This bamboo purse looks modern and appealing even though it is actually decades ago. It is no less beautiful than the famous products fashionable today.

212　**馬蹬一對**

Pair of stirrups　18×16×7 cm (2)

倒「U」字形的馬蹬是以厚竹片烘彎製成的。

These inverted U-shaped stirrups were made by heating and bending thick pieces of bamboo.

(1)

(2)

213　**香花籃**

Flower basket　19×19×12 cm

吊掛於室內，供放香花用的。

This type of basket was filled with fragrant flowers and hung inside the house.

214　**信插**

Letter holder　36×23×4 cm

以竹條穿成，供插放信箋用。

This letter holder was made of pierced strips of bamboo strung together.

215　**麵包籃**

Bread baskets　（1）31×25×7 cm

（2）24×18×7 cm　（3）20×15×6 cm

六〇年代外銷竹藝中的主力產品。

Baskets such as these were the main export products of the bamboo handicraft industry during the 1960s.

216　**糖果盤**

Candy dish　18×18×5 cm

編製細膩精緻，供放糖果或香花用。

This delicate and finely woven dish was used to hold candy and flowers.

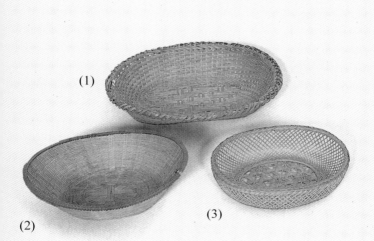

2
1
3
◄

2
1
4
◄

(1)

(2)

(3)

2
1
5
◄

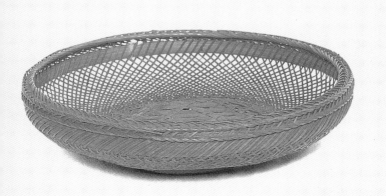

2
1
6
◄

217 **竹帽**

Bamboo hat 25×25×18 cm

原住民所戴的竹帽，堅硬厚實。

This type of hard, thick hat was worn by aborigines.

---

218 **竹帽**

Bamboo hat 30×24×15 cm

這是從前的警察或郵差騎腳踏車時所戴的。

This type of hat was formerly for police officers or
mailmen when they rode bicycles.

---

219 **火籠**

Heaters

（1）26×26×25 cm （2）16×16×16 cm

（3）15×15×15cm （4）9×9×7.5 cm

以前老一輩的人，在酷寒的冬天，幾乎是人手一籠的提
著火籠走到那兒烘到那兒，夜裡睡眠時還將火籠擺在床
沿取暖，在沒有電暖氣和空調的年代，火籠確實是冬天
裡的最佳良伴。

During the bitter winters, Taiwanese old people carried
these baskets everywhere they went to keep warm.At
night, they would place them next to the bed. In a time
when electric heaters and central heating were unknown,
these bamboo heaters were indispensable in the winter.

---

220 **麻將**

Majong 2.5×2 ×1.5 cm (144)

賭具。共144張牌。

Popular game. This set consists of 144 bone and
bamboo majong.

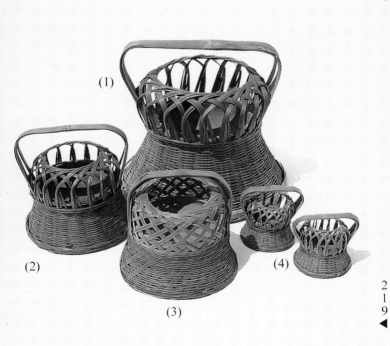

(1)

(2)

(3)

(4)

221 針線筯
Sewing basket 41×41×5 cm

從前沒有縫紉機的時代，所有針繡全賴手工，女子出嫁前都得練就一雙好手藝，因此針線筯是當時不可少的嫁粧。

Before sewing machines were used, all sewing and embroidery was done by hand. Each girl learned this skill before she got married and her dowry inevitably included a sewing basket.

---

222 針線筯
Sewing basket 42×42×5 cm

針線筯是以竹篾編內外雙層結構，內層多為兩種著色不同的竹篾所編成的整齊的圖案，中央處常編有「福」字或「卍」字，以示吉祥。

Sewing baskets are made of two woven layers. The interior is woven with two different colors to create a pattern in the middle of which fu characters or swastikas are commonly seen for good luck.

---

223 什細籃
Miscellany basket 34×34×13 cm

以竹篾編成，形似臉盆，外圍繪有墨畫裝飾，專供婦人裝針線衣物等細軟用。

This basket is woven with bamboo strips and is shaped like a washbasin. The outside has been decorated with ink. It was used by women to hold items such as sewing paraphernalia and clothing.

---

224 什細籃
Miscellany basket 37×37×13 cm

此件什細籃外圍雖無墨繪，但編者刻意以有去竹皮及無去竹皮的篾條間隔相編，呈現另一番美感。

Although this basket has not been decorated with ink, the artisan has interwoven strips of bamboo with skin and without to create a different but equally beautiful effect.

2
2
1
◀

2
2
2
◀

2
2
3
◀

2
2
4
◀

225　衣帽盒

Clothing and hat box　53×40×40 cm

以細竹篾編成，共二層，是從前放置官帽或閨秀擺放小孩的衣帽所用的。

This two-tiered box was woven of slender strips of bamboo. It was used to store the officials' hats or children's clothing.

圖225分層的特寫。

View of separate tiers of fig. 225.

226　竹筍

Bamboo chest　66×44×23 cm

即衣箱，供置放衣物或棉被，出外遠行時也可當行李箱用。

This chest was used to store clothing or blankets and could also be used as luggage when traveling.

227　竹筍

Bamboo chest　72×50×30 cm

此類竹筍，在出土的漢代文物中即有，形式相同，規格略小，證明竹編技藝的淵源流長。

Bamboo chests of similar shape and smaller size like this one have been found among excavated Han dynasty artifacts. It proves the long history of bamboo weaving.

228　書生籠

Book basket　45×45×45 cm

以前讀書人用以裝運書籍的竹籠，夾層中特別襯有棕櫚葉鞘，以防水滲。

Students used to carry their books in baskets such as this one long time ago. The layer of palm leaves inside helped to keep the basket dry.

229　什細籃

Miscellany basket　36×36×15 cm

這是編製手法較爲細緻的一件什細籃，樣子極好，少有。

This basket displays more refined artistry than the other kind. Miscellany basket of this style is less common.

230　儲物籠

Storage basket　47×47×35 cm

用來存放衣物家當或做行李箱用皆適宜。

This type of basket was used to store clothing and household items, or used as luggage.

231 竹簾

Bamboo blinds 248×128×3 cm

在台灣早期的民屋正廳中央，通常設有上下供桌，供奉祖先和神佛塑像或畫像，這件長超過八尺的竹簾，繪有天官賜福的神像，即是被供奉掛在廳堂中的，雖然年代久遠，所繪之礦物彩顏料部分已氧化或斑駁，但仍是早期台灣住民生活竹器中深具意義的佳品。

In the past, altar tables and portraits or carvings of ancestors and gods were set up in the main hall in Taiwanese house. This blind, with a portrait of a god giving blessings, would have been hung up for worship in the main hall of a house. Although some of the mineral pigments used to paint the blind oxidized or mottled, this excellent work is still greatly significant as an early Taiwanese artifact.

---

232 竹蓆

Bamboo mat 182×182×0.5 cm

此件竹蓆是由九隻長一百八十二公分，直徑七～八公分的長枝竹細剖穿成，平均每隻竹稈約莫剖成寬三釐米的細竹條八十條，整件竹蓆就是由這樣近七、八百條的細竹條穿孔，用麻線串成，製做工時浩大，尤其要將竹筒剖成粗細勻稱的竹條更非手藝老練的竹工不可，當今能有此藝製作者，已寥剩無幾，因此這樣看似簡單無奇的一件竹蓆，著實可貴。

This mat is made of nine branches of bamboo, measuring 182 cm long and with a diameter of seven to eight cm. Each branch has been cut into 80 strips about 0.3 cm wide so that the entire mat consists of 700 to 800 thin strips each pierced with a hole and strung together.

Producing such mat requires a lot of time and great skill to ensure that each strip is consistent in its thickness and width. Only a few artisans remain who are capable of this work. Therefore, what appears to be a simple, unassuming mat is actually quite precious.

2
3
1
◀

2
3
2
◀

參考書目

Bibliography

《台灣常民文物展——信仰與生活》
國立歷史博物館，1998。

《竹編工藝》
陳正之，國立傳統藝術中心籌備處，1998。

《台灣傳統工藝》
莊伯和，國立傳統藝術中心籌備處，1998。

《台灣懷舊之旅》
黃金財，時報文化出版企業有限公司，1998。

《台灣歷史圖說》
周婉窈，聯經出版事業公司，1997。

《民俗台灣》
林川夫編，武陵出版有限公司，1997。

《懷念老台灣》
許蒼澤‧康原，玉山社，1995。

《竹藝之美》
南投縣立文化中心，1991。

《台灣農業興衰四十年》
蕭國和，自立晚報社文化出版部，1987。

《竹書》
豐年社，1982。

《台灣早期民藝》
劉文三，雄獅圖書股份有限公司，1978。

《台灣民間藝術》
席德進，雄獅圖書股份有限公司，1974。

本書之付梓
特別感謝下列先進指導協助

台灣省政府顧問　簡榮聰先生

國立故宮博物院科技室技術指導顧問　彭定松先生

新加坡資深藝術收藏家　傅金洪先生

衣淑凡小姐

財團法人廣達電腦教育基金會　吳日曦先生

國家圖書館出版品預行編目資料

20 世紀台灣住民生活竹器＝20ᵗʰ CENTURY
TAIWANESE BAMBOO CRAFTS
／李贊壽著——初版——臺北市：藝術家，
民 89　面；公分——（　）
參考書目：面
ISBN　957-8273-72-X（平裝）
1. 竹工-圖錄
479.76025　　　　　　　　　　89011528

二十世紀台灣住民生活竹器

20ᵗʰ CENTURY TAIWANESE BAMBOO CRAFTS

李贊壽／著

發行人　何政廣

主　編　王庭玫

編　輯　郭冠英、江淑玲

美　編　李宜芳

攝　影　賴建作

英文翻譯　陳文心（Stacy Yu Chan）

出版者　藝術家出版社
台北市重慶南路一段 147 號 6 樓
TEL：（02）23719692~3
FAX：（02）23317096

總 經 銷　時報文化出版企業股份有限公司
桃園縣龜山鄉萬壽路二段351號
TEL：（02）2306-6842

台中縣潭子鄉大豐路三段 186 巷 6 弄 35 號
TEL：（04）5340234
FAX：（04）5331186
製　版　新豪華彩色製版印刷有限公司
印　刷　東海彩色印刷有限公司
初　版　2000 年 8 月 31 日
定　價　台幣 480 元

ISBN　957-8273-72-X

法律顧問　蕭雄淋
版權所有・不准翻印
行政院新聞局出版事業登記證局版台業字第 1749 號

贊助

財團法人|國家文化藝術|基金會
National Culture and Arts Foundation